MW01492964

ELEVATOR

and

ESCALATOR

RESCUE

Fire Engineering

ELEVATOR

and

ESCALATOR

RESCUE

A COMPREHENSIVE GUIDE

Theodore Lee Jarboe and John O'Donoghue

Disclaimer: The recommendations, advice, descriptions, and the methods in this book are presented solely for educational purposes. The author and publisher assume no liability whatsoever for any loss or damage that results from the use of any of the material in this book. Use of the material in this book is solely at the risk of the user.

Copyright © 2007 by
PennWell Corporation
1421 South Sheridan Road
Tulsa, Oklahoma 74112-6600 USA

800.752.9764
+1.918.831.9421
sales@pennwell.com
www.pennwellbooks.com
www.pennwell.com

Marketing Manager: Julie Simmons
National Account Executive: Francie Halcomb

Director: Mary McGee
Managing Editor: Jerry Naylis
Production: Sheila Brock
Production Editor: Tony Quinn
Book Designer: Susan E. Ormston Thompson
Cover Designer: Clark Bell

Library of Congress Cataloging-in-Publication Data

O'Donoghue, John, 1939-
  Elevator and escalator rescue : a comprehensive guide / John O'Donoghue and Ted Jarboe.
      p. cm.
  Includes bibliographical references and index.
  ISBN 978-1-59370-076-8 (hardcover)
  1. Elevators--Accidents. 2. Escalators--Accidents. 3. Elevators--Safety measures. 4. Escalators--
Safety measures. 5. Lifesaving at fires. 6. Emergency medicine. I. Jarboe, Theodore L. II. Title.
  TH9445.E48O36 2007
  628.9'2--dc22
                                                                                    2007003033

All rights reserved. No part of this book may be reproduced, stored in a retrieval system, or transcribed in any form or by any means, electronic or mechanical, including photocopying and recording, without the prior written permission of the publisher.

Printed in the United States of America

2 3 4 5 6   12 11 10 09 08

This book is dedicated to the firefighters who died on September 11, 2001, many of whom were trying to extricate trapped passengers from elevators.

Contents

Contents

Acknowledgments

This book has been a labor of love for both of us, and for many other people who were involved either directly or indirectly. First, we want to thank our families, our wives and children, brothers and sisters, and friends, without whose encouragement and patience this book would not have been published. We would be remiss if we didn't recognize the contributions of John's wife Beverly, for her invaluable editorial assistance, his daughter Caitlin, for her countless hours of typing drafts and supporting documents, and Ted's wife Cynthia, for her patience and support throughout writing of the book. They gave up more than we will ever know to allow us to complete our book.

We also want to thank Jim Gaut, Paul Lannielo, Robert Moore, and Robert Wilcox; and Kenneth W. Souza, Brandon M. Hugh, Mathew E. Ansello, and Mark N. Tiede (Rescue Company, Cambridge, Massachusetts Fire Department) for their assistance in taking pictures.

A group that played a critical role was the Peer Review Group (PRG), which consisted of Ed Donoghue (Edward A. Donoghue Associates Inc.), Matt Martin (Schindler Elevator Corporation), Doug Witham (GAL Manufacturing), Dave McCall (Otis Elevator-Canada), Charlie Murphy (Massachusetts-Department of Public Safety-Retired), Chief of Field Services John K. O'Donnell (Boston Fire Dept.), Assistant Chief John J. (Jack) Brown, Jr. (Loudoun County (VA) Department of Fire and Rescue Services). The PRG volunteered to review and comment on the chapters as they were written, and provided an irreplaceable level of expertise for which we are eternally grateful. It should be understood that although the PRG provided technical input, we authors are responsible for everything as written. The elevator industry responded to requests for pictures and other information that we needed to create the complete book, which we wanted to produce without hesitation or restrictions.

Other early supporters of helping the fire service teach firefighters proper safety procedures were companies who are now gone: F.S. Payne Elevator Company, with Bob Sheehan, the late Gahr Finney, Tom and Pat Coyle, and many others who always showed a helpful attitude. We thank Westinghouse Elevator Company, Dover Elevator Company, and countless small companies and individual mechanics who would take the time to show us both the way to do it right. Without their help, this book would not have been possible, and the companies today include: Kone Elevator, Otis Elevator, Peelle Elevator, ThyssenKrupp Elevator, Schindler Elevator, Fujitecamerica Elevator, G.A.L. Manufacturing, C.J. Anderson Co., Elevator and Escalator Safety Foundation, Adams Elevator, Edward A. Donoghue Associates, National Elevator Industry, Inc. (NEII), Motion Control Engineering, Carl White Associates, Smoke Guard, Inc., The Center to Protect Worker's Rights, Silver Spring, Maryland, and the American Society of Mechanical Engineers (ASME).

The foundation of the book is embedded in our more than 80 years of combined field experience, training, and sharing of information on elevator rescue operations. During most of those years, Ted's brother, James (Jim) E. Jarboe, Fire Chief, Takoma Park (MD) Volunteer Fire Department, and John E. Fisher, III, Firefighter, Ocean City (MD) Fire Company, worked with Ted in developing and delivering elevator rescue training within Montgomery County (MD), and in other regions and states. In Massachusetts, as director of the Department of Fire Services, Fire Marshal Stephen D. Coan, a great friend of the fire service, has played a critical supportive role in all elevator training and development. He continues to provide that support to this day.

Without the input and help from all of the preceding people and companies, we never could have put together what you are about to read. If we have unintentionally omitted any individual or company, we thank you for your help and apologize for any oversight. We also want to thank our editor at PennWell Publications, Jerry Naylis, whose advice and support were invaluable. The final aspect of the publishing puzzle is the input from the firefighters we have known, talked to, and worked with throughout our careers. You, brothers and sisters, are the true inspiration for this book. Read it and keep providing a safe operating environment for the public and the fire service.

Stay safe!

John and Ted

Introduction

Respond: To the Stalled Elevator With Entrapment

How often have we heard those words over the department radio or house dispatching systems over the years? When we first came to work in the 1960s, (yes, there were elevators even way back then), it still sent a quiver of fear through one's stomach, as we were quickly learning that we were going out into the world of the unknown practices of the fire service. With few exceptions, it was a gamble as to what we were going to do on arrival. If it was a good night, someone in the crew had a vague idea of what we were doing, but unfortunately, many times it was hit or miss, and we had no plan, other than get them out as fast as possible. We would watch as a lieutenant asked for a *metal* coat hanger, then twisted it out of shape and stuck it up into the void between the door and the metalwork framing of the elevator opening, trying to pick the interlock (600 volt AC). After 15 minutes or so, he'd give it up, and call for "Harry with the Halligan," who either quickly made short work of the interlock or just made everything impossible by completely trashing the door, but not getting it open. We both have been there— *power to elevator still on—no lockout tagout*—absolutely no safety considerations. What we, and you, have seen and done was to endanger not only ourselves and our fellow firefighters but also those who were safe until we arrived—the people inside the elevator.

Our careers were parallel to each other in many ways, even though we were hundreds of miles apart, with no knowledge of the other's paths and interests, until late in our careers. Both of us have spent much time and effort in developing, delivering, and pushing firefighter safety operations around elevators and escalators. As firefighters, you have the safety of the public in your hands, and that is why we have written this book, to provide you with the guidelines and information to avoid the costly errors, mistakes, and poor decisions that have cost people their lives and limbs.

Why Is the Fire Service Involved?

Many times while at an elevator call, we hear from the members, "What are we doing here? This isn't our job!" They are right in asking that question, but just as we respond to many types of calls in our daily responses, this type is one that has just grown on us due to the danger that others perceive the trapped occupants to be in, involving either a stalled elevator car or the more expected fire department response—that of a crush injury or an industrial accident of one type or another. On arrival, we are met with the request to assist those in the car, and that usually requires forcible entry tools and so forth. In that we are usually the only public safety entity that carries these tools, we have won the job. The firehouse adage is to take care of it now, or we'll be back in an hour to do it. This has resulted with the fire department responding to stalled elevator calls across the United States and Canada as part of our regular daily responses. If there is no emergency situation, and the occupants are just "locked in a box," they are safe. Call for the elevator mechanic, and whenever possible, await his or her arrival. In later chapters, we will point out the proper sequence for calling the proper elevator company to respond to your incident.

Now that we have established how we got here, we as the fire service, in particular the local fire department managers, have to make up our collective minds as to what role we are going to play. If we are not going to do it right and safely, it is best not to be involved at all. All of us have resented sitting in a training class, and having an instructor say, "You are either part of the solution or part of the problem." The problem is that they were usually right, and we can think of nothing that applies more closely than that of an elevator emergency. Either do it right, or don't bother becoming involved. Call for the elevator mechanic, or tell people not to call you for these nonemergencies, because by not being there, you will do no damage and endanger no lives. We all are familiar with horror stories of fire department responses to these incidents. Everyone meant well, but through a lack of training, procedures, and experience, members and the public were put into danger. A fire department responding to elevator calls without a strict Standard Operating Guide (SOG) is courting a disaster that is going to come back to haunt them and the public that they are dedicated to serve.

What Are We Dealing With?

As firefighters, we are always being exposed to the various types of building construction that are involved in a burning building. For some of us, that means buildings that could have been constructed during the 19th century or earlier, all the way up to the newest type of high-rise construction. Elevators are not any different when we look at the problems of existing and new construction elevators that impact our responses to elevator emergencies. The firefighter of today is dealing with working elevators that sometimes are 100 or more years old, particularly in some inner cities with mill-type construction buildings. In many cases, they are being operated despite the efforts of many to scrap them, due to the failure of local governments to mandate their retirement and replacement with modern units. The reason is the same one that drives everything that we interface with as firefighters—money! This is due to the vocal complaints and political pressure brought to bear by the business and building-owner community. Remember, if we are not present at the table, the politicians will vote to silence the squeaky wheel and pay us lip service about our wonderful dedication to the people.

It is important to remember that an older elevator is not required by code to be equipped with many of the safety features that we take for granted in newer units. In the United States there are two national consensus elevator codes: the Safety Code for Elevators and Escalators, ASME A17.1; and the Safety Code for Existing Elevators and Escalators, ASME 17.3. Every jurisdiction that has an elevator code has adopted ASME A17.1, though they may not be using the latest edition. However, the same is not true for ASME A17.3. Most jurisdictions just require existing installations to comply with the A17.1 portion of the code under which it was installed. A basic upgrade is required at various stages of an elevator's serviceable life span, when mandated changes are published and enforced, but it will not be a piece of equipment that is as good as it was when it was new. The elevator we all should fear is the one that has fallen through the cracks, whether it is deep in the inner city or located out in the country and has not been inspected or serviced in many years. This creature is just waiting for its chance to grab, kill, or seriously injure the unsuspecting person who might fall prey to its malfunction. If the elevator industry had their way, every elevator would comply with the most current codes and would be regularly inspected. That is the ideal world for both of us, but it is not the real world. We as firefighters must be ready to face the broad spectrum of the various makes and models of elevators that have been part and parcel of the elevator industry over this time period.

Who and What Is ASME A17.1 and CSA B44?

ASME A17.1 and CSA B44 are the organizations that publish the harmonized elevator codes for both the United States and Canada. ASME A17.1 is the Safety Code for Elevators and Escalators, and CSA B44 is the Safety Code for Elevators. Both were developed and are maintained by the American Society of Mechanical Engineers, A17 Standards Committee, in New York City. A new edition of the code is published every 3 years, with an appendix of changes published in the 2 intervening years. Those revisions come from the ASME A17.1 Standards Committee (35 members minimum), which meets three times a year across the continent, who approve the changes and corrections that the 25 subcommittees prepare. Revisions are usually based on recommendations from individuals, inspectors, other interested parties, and subcommittee members. Changes get bounced back and forth until consensus is reached by the ASME Standard Committee and the CSA B44 Technical Committee, then the changes become part of the standard. One might ask, "What does this have to do with us as firefighters responding to an emergency?" The answer is that they write the standard that affects the elevators that we ride and, depending on the situation, respond to in an emergency. The following is a listing of the ASME A17 Standards that affect us:

- ASME A17.1 Safety Code for Elevators and Escalators

- ASME A17.2 Guide for the Inspection of Elevators, Escalators and Moving Walks

- ASME A17.3 Safety Code for Existing Elevators and Escalators

- ASME A17.4 Guide for Emergency Personnel

Of primary importance is the ASME A17.1 Standards Committee subcommittee that is called the Emergency Operations Committee. This subcommittee meets three to four times a year around the United States and Canada, and it is responsible for the code requirements that define how firefighters' service Phase I and Phase II elevators will operate. This committee is made up of firefighters, inspectors, elevator system designers, and company representatives. The ASME A17.1 Emergency Operations Committee is also responsible for interpreting the requirements for Phase I and Phase II, subject to the responses being approved by the ASME A17 Standards Committee. Recent interpretations can be located on the ASME A17 Committee Web site. See the ASME Web site at www.asme.org.

Why Don't They Ever Listen to Us?

How often have we heard this lament at the firehouse kitchen table, concerning those who make the changes that affect our daily lives? The reason is this: We are seldom at *their* table! It is that simple, as most fire departments do not send members to meetings concerning the elevator codes or standards. Think about it for a moment: How seriously do we take the presence of someone at a fire service meeting, who has shown up only to put in their 10 cents worth when we know they will not return until the next blue moon? It is the same with elevators and escalators. We as a service do not usually attend the monthly or weekly meetings of the

local elevator industry. That is where all code changes start. It is also where you can develop the contacts you will need to help you maneuver the pitfall-filled route to code changes that we may want to pursue. Beyond the code changes, these meetings can be a valuable source of information about necessary equipment that you may need for training and for the development of career-long cooperation between the fire department and the local elevator industry members. At the first meeting that was attended by a firefighter as a member of the ASME A17.1 Emergency Operations Committee, the committee's members were very friendly and pleased that there was finally a firefighter in the membership. At the second meeting, a member took him aside during a break, to let him know his brother was a battalion chief in a large eastern city. He knew at that moment that he was not going to be alone at these meetings. Organizations such as the International Fire Service Training Association (IFSTA), National Fire Protection Association (NFPA), International Code Council (ICC), and other leading training standards developers and book publishers often quote the ASME standard in their manuals and standards, but they are not firefighters. If we do not provide the necessary input that is needed by those making changes, then they will take what others provide and that probably is not going to be in our best interest. We have found the members of the elevator industry only too willing to help us, but they need our support at these associations and organizations at the local level. Remember this: They all have a brother, sister, neighbor, friend, or brother-in-law who is a firefighter, and they want to assist us, but cannot do it in a vacuum.

Who Are You Going to Call?

When calling an elevator company for assistance at a stalled elevator call, there are certain ground rules that we have to understand and work within. First is that the elevator manufacturer (e.g., Otis, Dover, Payne) whose name is on the equipment may not have touched that unit in 10 years and may no longer even be in business (e.g., Dover, Payne). Your request for assistance may fall on deaf ears. When a unit is constructed, the company who built it usually has a service contract for a one-year period. When that period expires, the service companies will try to wrestle the buildings maintenance contract away by offering a deal for a very competitive price.

Now, that brings us full circle to the question of who you are calling to assist you at the incident. The A17.1 code requires that a 24-hour service card be located in the machine room for that elevator system. It should be there, but in many instances it may not be. On entering the building, you may be able to determine who is responsible for the repair by asking tenants, building engineering staff, or other sources on hand or by calling the elevator inspector's office. The bottom line is that you must develop a system of identifying the right company to be called for a response, and that must be coordinated with your fire alarm or 911 dispatch center. In chapter 11 (Safety) we will discuss the benefit and the procedure for a better response by the industry.

A very important point that should be discussed is the present state of the elevator industry in North America. When we first started to read elevator magazines (*Elevator World* is their version of *Fire Engineering*), there were 252 elevator companies in the United States that built, installed, and serviced elevators. Today the playing field has changed drastically, with the swallowing up of independent firms by multinational companies. Today there is only one multinational elevator company based in the United States, and that is Otis Elevator Company. Other major companies involved are Schindler Elevator Company (Swiss), Fujitec America, Inc. (Japan), KONE Corporation (Finland), and ThyssenKrupp AG (Germany), all of whom have manufacturing facilities in North America for some components while importing other components from elsewhere.

In this introduction, we have tried to cover *who, what, when, where,* and *why* of this problem. Ideally, in the succeeding chapters we can impart the *how* to you, the reader.

Chapter 1
The History of Elevators

The Past, the Present, and the Future

The past

As firefighters, we always study the age of everything that we come into contact with in our daily operations. We work in old and new buildings and have seen everything in between. The world of elevators is no different, and they have been around much longer than any of us would believe.

Historians speak of the pyramids of Egypt with awe when they contemplate how those early engineers moved large slabs of granite into position. Certainly, much of the credit can be ascribed to the labor of countless humans, but some of it is due to the leverage, pulleys, and ropes that moved materials vertically from point A to point B. This is the most basic purpose of an elevator.

As fire science students, we all learned about the very first organized fire department, which was in early Rome. A force of Roman centurion soldiers was reserved for fire-fighting activity, because of the thousands of fires used for cooking and other purposes in ancient Rome.

Rome played another early role for elevators. During the recent movie *The Gladiator*, starring Russell Crowe, he and other gladiators are herded into a large wooden box (the elevator) and are then lifted to the arena floor as slaves turn a wooden wheel device (the motor), which is actually a pulley system attached to a large overhead rope (the traction rope). Although this may be an oversimplification, it is also the mechanism of the elevator in the Roman Coliseum. Even today, the construction alcove that was used for the elevator hoistway can still be seen, as reported by James Goodwin in his book *Otis: Giving Rise to the Modern City*.[1]

The use of equipment to move material advanced the development of what we now call an elevator. The Industrial Revolution was the driving force that propelled many of the necessary step-by-step improvements that allowed a system that only moved materials into one that moved people. The moving of people was not without great risk, as the hemp rope that was commonly used for lifting parted on occasion, causing the platform with people and stock to crash to the ground.

The defining moment in the development of a safe means to transport humans came about in 1853 during the New York Crystal Palace exposition, when Elisha Graves Otis presented his "Safety Hoist" invention to prevent an elevator from falling (fig. 1–1). It consisted of notched hardwood oak rails and a cart spring, which dug into the notched wooden rail when the hoist rope separated. His invention, although simple when looked at relative to today's car safeties, is the father of them all. Without it, the progress toward having a safe means to transport people would have been delayed until much later.

Fig.1–1 Elisha Otis presents his "Safety Hoist" invention in 1853. (Courtesy of Otis Elevator Co.)

During the Industrial Revolution, from the late 18th century and continuing well into the 19th century, many different methods were used to power the movement of elevator platforms and the passenger elevator. These included the belt drive system, common to the textile industry and mill-type buildings, where all moving equipment was powered by a common belt drive into which all the other belts were tied in. Others were powered by steam, water hydraulic systems (still found working today), electric traction, and, later, oil hydraulics. All of these were used in various industries to move goods and, ultimately, people.

As the progress continued, many companies came into the market to compete for a share of the business. This competition brought about changes, improvements, and the development of new products that have continued to advance the industry. Competition is truly the great engine of development, as each company has its own designers and engineers working on tomorrow's products today. If they do not, then they will wither on the vine and disappear from the marketplace.

After the Industrial Revolution came to a close by the late 19th and early 20th centuries, elevators were progressively changing into the machines that we have all grown accustomed to seeing in our daily lives. Most of us remember seeing birdcage-type elevators, located in open hoistways, with an elevator operator asking, "Floor please, watch your step." The automatic passenger elevator has replaced the elevator operator, where decisions about what floor are left up to the individual. The change from operator driven to automatic operation revolutionized the elevator industry.

The present

Although it may sound like ancient history to some of our readers, the advent of a period of great change in elevators came about directly from World War II. Wartime research and development spurred great technological changes. The early computer, metallurgical advances, and motor design and longevity all had their roots in wartime efforts. The advances in construction techniques that engendered taller and bigger buildings prompted the development of newer techniques for transporting people and freight. The number of hydraulic elevator installations went off the charts, as this seemingly simple system gained popularity everywhere.

The necessary building boom brought about by the reconstruction of most of Europe after World War II offered the opportunity to put these changes into place in the new buildings of companies that were forming, in competition with some of the long-established firms. Throughout the entire system, technological advancements have continued, from electromechanical contact controllers to today's solid-state computer-driven systems, which can be seen in every modern machine room today.

Whereas machine rooms once were very cold, dark places, with huge, ungainly pieces of machinery, the story is different today. The machine room of a system placed into service today is one that is well lighted, climate controlled, and very quiet. The control cabinets are covered and more than likely locked to prevent dust and grime from getting into the very fragile equipment. When elevator mechanics come to troubleshoot today, they use diagnostic software, leaving the bag of big wrenches out in the truck for another day.

The elevator of today is faster and safer than ever before, streamlined, and very much a part of a building's identity. Light-emitting diode (LED) panels provide stock updates, sports information, and quick access to emergency response in the event that a car stops, trapping the passengers. Car speeds have been gradually increasing, with the high-speed cars in the Empire State Building (1931) operating at around 300 meters per minute (m/min)(nearly 1,000 feet per minute [ft/min]). The car speeds in the Sears Tower in Chicago (1974) increased to the astounding speed of 488 m/min (1,600 ft/min). This pathway of development, design, and implementation leads to newer concepts and techniques, all resulting in change. The September/2005 issue of *Elevator World* reported that the world's tallest building (101 stories), Taipei 101 (2004), has two high-speed elevators capable of traveling at 1,010 m/min (about 3,300 ft/min).[2] The cars have an automatic atmospheric control system to assure comfort for the occupants as they travel in these "aerodynamic capsules." They can travel from the ground floor to the 89th floor in 39 seconds!

What might be the largest area of change that is happening is the expansion of the machine room-less (MRL) elevator onto the entire elevator arena. Although commonplace in Europe for the past 15 years, the MRL is just starting to have an impact in North America. When first introduced by individual companies, they were limited in height, usually 8 to 10 floors. Today, manufacturers are installing the MRL machine-in-hoistway equipment in low and mid rise buildings, and there is no doubt MRL equipment will be entering the high-rise market as well.

The future

The future of elevators is only limited by one's imagination and the physical barriers presented by the strength of the materials used in construction. Over the years, there have been many new entries into the market, only to find that they did not meet the expectations of either the manufacturer or the customer in whose building they now exist. It is not unusual to overhear elevator people talking about a product or design that turned out to be a nightmare for the owner and the service company that serviced it. If you buy a bad pair of shoes, you either throw them away or only wear them when necessary. When building owners buy into a new design or concept, they anticipate being on the cutting edge of technology and that it will be an added attraction to draw tenants to their building. The new designs are not just thrown into the market; rather, prototypes are tested through a mind-numbing series of tests in company test towers, where wear and tear can be simulated by actually conducting thousands of trips up and down the hoistway.

What is on the horizon in elevator technology? This can be broken down into two areas:

- The future designs and concepts for vertical and horizontal transportation.

- The future use of the same, particularly during fires and other emergencies.

The first area encompasses future designs, which are on the computers already via computer-aided design (CAD) systems. The term "on the drawing board" just doesn't exist any longer. Systems can be conceptualized, drawn, and actually simulated via the CAD systems that all designers have access to today. Elevators moving independently from one hoistway to another, much as a "Pac-man" image moves on a screen; multiple elevators using the same hoistway; cars moving via impulse motors affixed to all sides, using no hoist ropes or using alternative materials—all are part of the future. Concepts that we have not even heard about will be in tomorrow's television or newspaper article announcing another breakthrough in science and technology.

The second area dealing with the future use of elevators is one that must involve the fire service as an integral part of the design team. If we are not actively participating in the decisions being made about the future use of elevators during fires and other emergencies, we will find that others have made the decisions for us. We will read the newspaper at the kitchen table and find out that non-fire-safety experts have taken control of the elevators during a fire.

The fire service "owns" the elevators once Phase I has been activated, either manually or by activation of automatic recall. The incident commander (IC) has the legal as well as the moral responsibility to make decisions about the occupants of the building and what their actions should be. There are groups active in the access and evacuation fields who will, if allowed, take Phase I away from the firefighters, leaving the firefighters with a *few* elevators while they (in practice, a computer program) control the others for an evacuation of the affected floors. At present, a task group from ASME A17.1/CSA B44 is examining this entire problem in detail. This examination also includes the future of the "hardened elevator," which would allow the Phase II operation to continue through some conditions that currently would cause the cars to fail. Included will be comparisons of the European versus American concept of firefighter elevator operations.

This group is meeting three times a year to try to sort out the future distribution of the concepts to the responsible agencies across a wide spectrum of interested parties, including code and standard makers. The future of elevator design is in their hands, and the future use of elevators during fires and other emergencies must include the fire service.

Summary

As we have learned in this chapter, elevators have been around for a very long time. Both the Egyptian and the Roman Empires used them in one form or another. Their development continued through the Industrial Revolution, into the 19th and 20th centuries, and has never stopped. World War II provided a great impetus for both the development of components used in that industry, as well as the opportunity after the war for the implementation of that knowledge during the postwar building boom. Today's machine room, if there is one, is quieter, cleaner, and more comfortable than its predecessors in the past. The solid state electronics of

the system demand a different environment to support those components.

The future will be one of great changes, with systems operating that we cannot even imagine today. We, as the fire service, must be on the watch for changes in the use of Phase I in buildings during a fire emergency. There are those who want to take over the elevators for their own use during these incidents, and we cannot allow it.

Review Questions

1. Where in Rome was an early version of the elevator used?

2. What modern time period is considered to be critical for elevator development?

3. What year did Otis exhibit his Safety Hoist invention?

4. What does the second area dealing with elevators consist of?

5. What is the name of the tallest building in the world?

Field Exercise

Visit the oldest commercial building in your district and catalog the elevator by age, type, make, speed, and distance traveled.

Endnotes

[1] Goodwin, J. *Otis: Giving Rise to the Modern City.* Chicago: Ivan R. Dee, 2001.

[2] Gray, L. "A Brief History of Residential Elevators: Part 1." *Elevator World* (January 2005), p. 110.

Chapter 2
Elevator System Overview

Fig.2–1. "Down the hatch!"

The Hoistway

Fig.2–2. Looking down into the hoistway

Even though we are all familiar with this expression, it means something totally different when we are dealing with elevators (fig. 2–1). The elevator hatch, or more properly, the *hoistway,* is the area where the elevator "lives." It is its domain, its lair, and those of us who enter it without the proper respect for the danger that lies within will pay the price.

The elevator hoistway in today's elevator system is usually constructed to a 2-hour fire resistance rating (fig. 2–2). The older versions will vary considerably: there will be wooden guide rails and car components such as flooring, side walls, poorly or totally unprotected floor openings for doors, windows, and the like. The types of construction will vary from wooden buildings, solid brick, cement block, and composite (gypsum) board construction, to glass enclosed or no enclosures on observation elevators installed in atriums and on the outside of buildings. Today's modern hoistway will be vented to move smoke and heat out to the atmosphere, and may be pressurized to keep the products of a fire from entering it from a fire floor.

The *hoist ropes* enter the hoistway from the machine position, which may be above, below, or to the side, and descend down to the car top, where the elevator is attached to its set of guide rails. These ropes are made of twisted steel wire, and during a fire, they will melt and deform, eventually separating due to the high heat that will accumulate in the hoistway. Remember that the hoistway is the best chimney in the building, with a great potential for smoke and heat to spread to the upper floors of the fire building. The *guide rails* are attached to the walls of the hoistway by a series of brackets, designed to provide the steadiness needed by a moving system such as an elevator (fig. 2–3). The cars are attached via sets of *roller or sliding guides,* which may be either rigid, swivel, or roller type, which will allow the fast smooth ride that we are used to today (fig. 2–4). There are two sets of guides attached at the top (crosshead beam), and two sets attached at the bottom (car safety plank) of the elevator (figs. 2–5, 2–6, and 2–7).

Fig. 2–3. Roller guide shoe

Fig. 2–4. Slide guide

Fig. 2–5. Car hoist rope attachment

Fig. 2–6. Gen2 belt attachment

Fig 2–7. Hoist rope construction

As the hoist ropes reach the car, they are attached at the crosshead beam using shackles. On the top of the car are other parts of the system, namely, the *car top inspection station, car top emergency exit,* and the *car door operator* (fig. 2–8). Also note the location of the *counterweights,* which we will discuss later. All of these play an important role in our activities during an elevator emergency. The car door operator is the motor that provides the means that the elevator uses to open and close the car door and hoistway door when at a landing. The car top emergency exit is to provide a means of access for trained personnel to facilitate the removal of passengers when

the conditions warrant. Depending on the age of the elevator and whether it is in a Seismic Zone 2 or greater (see ASME A17.1 section 8.4.4.1), the emergency exit may be locked from the outside, with a safety circuit contact. The safety circuit contact is not on all elevators built in the past, due to different requirements in various editions of the ASME A17.1 Safety Code. Others have more basic obstructions, such as wing nuts, screws, nuts and bolts, slide bolts or, in some jurisdictions, the Phase II key. Unfortunately, some have no means to prevent children and students and the foolish from gaining access to the top of the moving car. The car top inspection station, although an inspection device for elevator mechanics, does provide access to an emergency stop switch (red), and a light switch (white) is provided to help see in this dark environment. The modern elevator has a prescribed process that must be followed to have the components of the station activated, because it is for mechanics and inspectors only, not available to us. A firefighter may push the up or down button located on this station, but these features are not "alive" at that time. Only the emergency stop and the light switches are active at all times.

Guide Rails and Governor System

Fig. 2–8. Car top inspection station

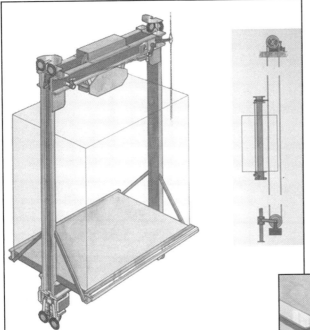

Fig. 2–9. Car sling

The counterweight is equal to the weight of the car plus 40% of its load-carrying capacity. When an elevator system has a malfunction with its braking system, the car will usually "fall up," striking the ceiling of the hoistway, not the pit as movies would have us believe. Remember, the next time a member suggests moving an elevator by opening the brake, that the car is coming up the hoistway, not going down! Many members of the elevator industry can recount harrowing tales of near misses or show hands with missing fingers brought about by forgetting to be wary of the counterweight of the elevator they were working on, as well as the one belonging to the adjacent car. Both of these sets of guide rails descend the length of the hoistway into the elevator pit.

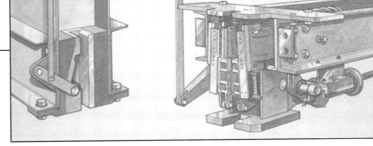

Fig. 2–10. Car safety plank

The sides of the cab are of composite construction, depending on the grade of the system. The sides of the sling of a cab are called the *stiles*, while the remaining part of the car sling is called the *car safety plank*, and is usually the strongest and biggest part of the system (fig. 2–9). The reason for this is that the *car safeties* (gradual or instantaneous) are attached to this heavy steel member due to the stresses that will be placed on it during the operation of the car safeties when an emergency stop is actuated by the governor system (fig. 2–10). While familiarizing ourselves with the hoistway, we should make note of another set of guide rails in addition to the ones that our car rides on. This is the counterweight set of guide rails, and may be located on the side, behind, or in between the elevators of our system. Counterweights are a very dangerous piece of the puzzle, as they move silently in tandem with the elevator, except that they are moving in the opposite direction, and may catch the unaware member of the fire service by surprise with deadly results (fig. 2–11).

Car Sling and Car Construction

Fig. 2–11. Counterweight with Otis belts

In the pit, in addition to the anchor points for the guide rails will be the governor tension sheave wheel, limit switches of various kinds, spring or hydraulic buffers, and other pieces of necessary equipment for the system to operate. As the car ascends and descends, a cable will be seen, hanging from under the car, following it up and down the hoistway. This is the halfway line (also known as the traveling cable), and it comes out of the wall at the halfway point in the hoistway wall (fig. 2–12).

Fig. 2–12. Halfway line (traveling cable)

This carries the 110 AC power for the car lights, fan, and so forth. At this point, you may also see the *compensation rope* or cables, and the like. Their purpose is to partially or fully compensate for the weight of the suspension ropes or belts. This compensation means must be fastened structurally to the counterweight and the car. Every elevator system must have an emergency stop switch and a light switch located inside the hoistway, on the wall on doorjamb side, at the lowest and uppermost access to the hoistway. The entrance to the pit is either a *jump* or *door pit*, with access as the term indicates. If there is a door into the pit, it will be at the lowest access to the hoistway and can be opened by a service company key. It is monitored by a safety circuit contact that will apply the brake in the machine room when the door is opened. This is to provide protection to the mechanics

when they have to enter these areas. In the pit and on the top of all elevators there has to be an area of refuge, which serves as the place for a mechanic to move into in the event of an unanticipated movement of the elevator (fig. 2–13).

Elevator Car Safety Devices

Fig. 2–13. Car top refuge area

A firefighter would never find this spot in an emergency, as every model and make is different and requires a working knowledge of the system to take advantage of it. Also located in the pit are the buffers, either of the spring or hydraulic types. They are located on pit channel irons, to help distribute the anticipated load that may come on them (fig. 2–14). The best plan is never to be in the pit or on top of an elevator that has power still to it. *Power off, and lockout/tagout is your only true protection.*

Pit Contents

Fig. 2–14. Elevator pit equipment

The elevator hoistway and its equipment is a place that takes many years of training and experience to become proficient to work within safely. In this chapter, we have taken you into the machine room, and showed you the contents of it and the hoistway, the basic construction of the elevator car, and the facilities in the pit. As firefighters, we are only visiting this workplace of others either as an inconvenience or in the event of a true emergency. We should always remember who and what we are, and that we are not elevator mechanics, but firefighters, and that this is a very dangerous and deadly place. Keep that in mind, and everyone in the company should return to quarters safely.

Summary

The basic construction of an elevator starts with the hoist ropes descending to the car from the machine room that controls it. The ropes, as they are called,

attach at the top of the car in an area referred to as the shackles. It is here that they then are attached to the crosshead beam. The side pieces of the construction are called stiles. The bottom is called the car safety plank, and this is the strongest part of the elevator system. Later, the car safeties will be attached to this beam, and when activated will exert tremendous force onto that beam. This structure that all of the pieces complete is referred to as the sling.

The car itself runs on its own set of guide rails that are attached to the walls by brackets. The car runs on either roller guide wheel sets or slide guides, which allow the smooth ride that we are used to.

On the top of the elevator are the following:

- The car door operator motor
- The top of car emergency exit
- Inspection station for the mechanic to use

Finally, the examination of the system is completed by visiting the pit. The various pieces of pit equipment include the counterweight screen, buffers, governor-tightening sheave wheel, guide rails fastenings, and other necessary equipment.

Review Questions

1. What is the hourly fire resistance rating of an elevator hoistway?
2. Define the role that the car door operator plays in opening the elevator door.
3. How do the car safeties operate?
4. What is the other name for the halfway line?
5. Compare a slide guide to a roller guide.

Field Exercise

Set up visits by all of the companies and groups to two distinctly different elevator installations in your area. One should be a modern (5 years old or less) installation, and the second should be at least 50 years old. Use your local authority having jurisdiction (AHJ) to help locate these installations.

Chapter 3
Traction Elevators

The elevator machine room as we knew it has seen a great number of changes in the past 30 years. The "old clunker in the dungeon," as we knew it, still exists, with machines running that look like they came out of the dark ages. *(They did!)* In the past, we would encounter machine rooms (MRs) with wide-open electrical boards (controllers), grated openings in the floor that looked down into the hoistway (hoistway vent exchange), huge geared and gearless machines, poor lighting, and no heat or air conditioning (figs. 3–1 and 3–2). The weather was at times clearly a problem, as water would be our constant companion on the floor.

Machine Room Contents

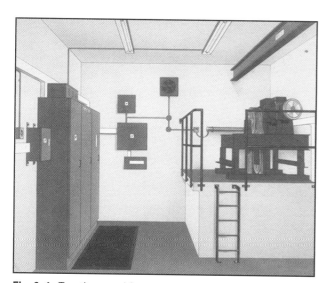

Fig. 3–1. Traction machine room

Fig.3–2. Old relay logic controller

11

A trip across the roof in January after an ice storm provided one with an adventure even before getting to the machine room door.

Today, the modern machine room has many features that make it a more hospitable place, but certainly just as dangerous to the uninitiated firefighter walking in the door. The location of the machine room can differ from one family or type of elevator to another. The hydraulic machine room is typically in the basement or lowest floor of the system, but may be anywhere they can put it to save the rentable floor space. The traction system machine room may be located in the penthouse, roof area, or the basement.

The newest family of elevators, the machine room-less elevators (MRLs; see chapter 9, New Technology) may be in a control space, cabinet, or room adjacent to the hoistway at the top or lowest floor, or may have all components within the doorjamb or hoistway wall, depending on the design of the particular system. To open the door to the modern machine room, you will need a key, usually an elevator service company key that the service company and the building manager should have (fig. 3–3). The fire department that responded to this incident should have brought the "Big 8" in with them when they stepped off of the apparatus.

These include the following items:

• Portable radio

• Flat head axe

• Halligan tool

• Hand light

• Lockout/tagout set

• Hoistway door unlocking device key (hoistway door key)

• Tool for poling

• Hydraulic ram tool (e.g., rabbit tool)

These tools will be necessary if forcible entry is required into the machine room, if a key is not available in the Knox-Box or a similar receptacle or from building management. A locked machine room door is never an excuse for not getting into the room and performing a *power down* sequence, by operating the mainline disconnect switch to the off position, and locking/tagging it out (fig. 3–4). When opening the door, you will find the power disconnect switches or panel in some jurisdictions located on the door jam side within 18 inches of reach, and at 5 feet 0 inches in height on the wall. Unfortunately, the National Electrical Code (NEC), whose responsibility it is, does not locate them as strictly, and you may have to look around to find them. The firefighter will also notice a separate set of disconnects or lockable electrical switches that control the 110-volt power for the elevator. It is *not* advisable to shut these off, as they control the lights, fan, and the like, in the elevator. It is bad enough that the occupants are in the stalled car, but we do not have to add insult to injury by also turning their lights out!

Fig. 3–3. Machine room with signage

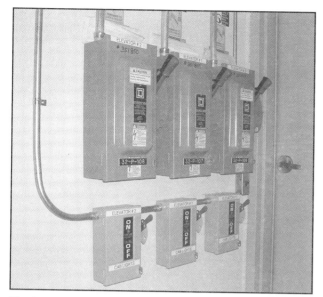

Fig. 3–4. Well-marked mainline and 110-volt disconnects

At this location will also be found the shunt trip device either inside the machine room or outside of it on a wall. Whatever the shunt trip's location, the mainline disconnect itself will always be in the machine room. It will disconnect the mainline automatically before water is allowed to flow from the machine room sprinkler. (This is discussed in length in chapter 22.)

In older rooms, you may find the mainline disconnect located anywhere in the room that met the elevator code of that era. The biggest change that you will notice is that the machine room is now air-conditioned (50°F to 90°F) to accommodate the needs of the solid state electrical components that require a comfortable working temperature. The solid-state phase of elevator design brings into focus the maximum temperature ratings for the commercial chips used by the industry. Commercial chips are rated to 70°C (158°F) (mostly), with military versions going higher (100°C/212°F to 140°C/284°F), depending on the chip. If the machine room gets too hot, the elevator control system may fail, possibly trapping people in the car. The exact temperature limit varies from elevator to elevator. The system may recover when the temperature is lowered, or it might need the attention of an elevator mechanic.

An important point is the electrical capacity allowed in the elevator machine room and the hoistway. It is allowed to reach up to 600 volts AC, which is certainly enough to get your attention. The modern machines are smaller and quieter, but still just as dangerous as the older ones.

The only action that a firefighter needs to take once inside the room is to select the appropriate disconnect switch, identified by its car number, and pull the lever to the off position. According to ASME A17.1, rule 2.29.1, the number must be located in numbers two inches high on the mainline disconnect, controller, and motor. The current ASME code requires the car number also to be displayed on the car operating panel (COP) in the car. After performing lockout/tagout, you must stand by the room to prevent anyone from removing the lockout/tagout until your members are safely finished with their actions. Also, be aware that some older mainline disconnect electrical boxes cannot be locked out, and a physical presence may be required to assure that after shutting off power, no one prematurely reenergizes the system while members are operating. Always leave the power disconnect in the locked-out position until an elevator mechanic can make a determination that the elevator is safe to use. If this means having to go back to the scene later to remove your equipment, then that is what must be done.

Our role is to safeguard the public, and that is what we are there for. It is important to remember that the only function that we have in the machine room is to perform power down and lockout/tagout, and stand by in the hall, away from other moving equipment. Maintaining a presence may be beyond the scope of lockout/tagout, when staffing permits, it is highly recommended. Keep in mind that our turnout gear and tools provide an excellent opportunity to ground us to high electrical sources (electrocution) or have our clothing caught by moving equipment (traumatic amputation) that clearly does not give you a second chance.

An even more critical point that must be covered is that under no conditions does the fire service ever open or operate the brake of a stalled elevator. Once the brake is released, normally the car is moving *up* the hoistway, because the weight of the counterweight is equal to the weight of the car and any accessories, *plus* 40% of its rated capacity.[1] If the car weighs more than the counterweight due to overloading, then the car will move down the hoistway. *This is an action that must only be done by an elevator mechanic, who is trained to perform this function and will know when the situation warrants its use.* The photo at the beginning of this chapter shows a general impression of a machine room. Now, we should become specific, and cover all of the entities that we will find, and in what order, as we enter the room.

Power Disconnects

As stated earlier, power disconnects are located on the wall inside the machine room. Distinguish between the drive power and the 110 power, and when ready, shut off the drive power and perform lockout/tagout according to the proper procedure. It is important to maintain a Fire Department presence throughout the incident, to prevent someone coming in, removing your equipment, and endangering the members and the public.

Traction Machine Room

The type of machines in these rooms may be either the drum, geared, or gearless varieties (fig. 3–5). Simply, the *gearless* (typically 450 feet per minute [ft/min] or greater) is driven by a large motor, with the motor, drive sheave, and brake on a common shaft (fig. 3–6). They are usually installed in mid- and high-rise buildings, but are now being found frequently in low-rise installations as a result of the drop in price and easier maintenance. The *geared* (typically 450 ft/min or less) uses a worm and gear wheel setup to accomplish the power transfer from the motor to the drive sheave (fig. 3–7). This can be compared to the power take-off (PTO) that we use when, for example, Engine #6 engages its pump while operating at a fire. The location of the machine room for a traction machine is usually directly above the hoistway. This is due to physical necessity, because the ropes must be able to create their traction (lift) in a direct manner, via the traction sheave wheel on the traction machine (fig. 3–8). On one side of the machine, the ropes go to the car, and on the other side they connect to the counterweights (normally in the same hoistway). In some really ancient rooms, ropes actually come across the room to accomplish an engineering requirement of one kind or another.

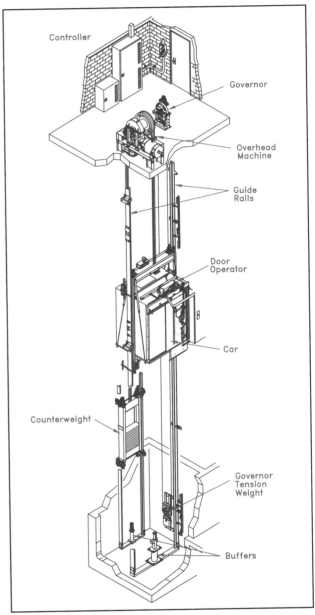

Fig. 3–5. Traction schematic (courtesy of KONE Corp.)

Fig. 3–6. Gearless machine

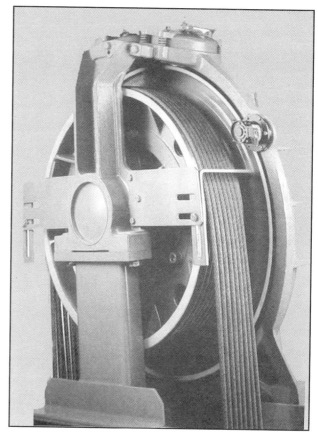

Fig 3–8. KONE PM traction

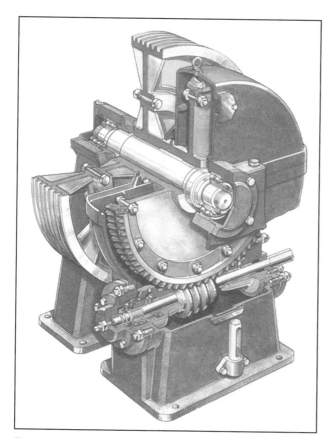

Fig. 3–7. Geared machine

In close proximity to the machine, you will find a governor. Its function is to monitor the speed of the car, and apply a car safety if it detects a speed higher than a preset value. The governor is driven by a rope that is separate from the traction ropes. When an overspeed condition is detected, a set of jaws on the governor grip its rope and set the undercar safeties, which will stop the car. Since the ASME A17.1–2000 edition of the code, protection must be provided against overspeed in both the up and down direction. The distance necessary to stop the car is a function of car speed, capacity, and type of safety.

Across the machine room are the controller cabinet(s), which contain the brains of the elevator operation (fig. 3–9). In a modern room, you may also notice a laptop computer station, which allows the mechanic to selectively troubleshoot the controller program that runs each car. These controller cabinets are a source of electrocution, and we have no need to go anywhere near them (fig. 3–10). Since we are not training to be mechanics, it is not necessary for us to go into detailed explanations of dimensions, widths, sizes, and the like. What we as fire fighters must know are the necessary and proper functions that we can accomplish safely and without endangering ourselves or the public.

Fig. 3–9. Solid-state controllers

Fig. 3–10. MCE gearless machines and controllers © 2005 Motion Control Engineering Inc.

Hoistway Vent Exchange Opening

In older rooms, we will still see the grated vent openings (never *stand on it!*) that cover the hoistway. There are many reasons for their existence, such as cooling and heat exchange via the piston effect of the elevator's movement or to enable mechanics to view a car as they are moving it. What we do know is that, during a fire in a building, they provide to the products of fire (smoke, heat, and toxic gases) a clear and unobstructed pathway via the best chimney in the building, the hoistway, into the machine room. Here the heat and smoke attack the controller and knock the elevators out of commission very early in a fire. Unfortunately, in some cases they have severely contributed to smoke mushrooming onto upper floors when they were covered over for the winter to keep the machine room warmer. Today's hoistway is required by the building codes to be vented below the machine room to the outside of the building (figs. 3–11 and 3–12). Those same codes have also limited the size of openings for the ropes, governor, and so forth, which would allow products of combustion access into the newer room. A big difference is that today's controller will have its doors shut and locked, unlike its predecessor, which we found with its doors either wide open or nonexistent.

Fig. 3–11. Hoistway vent exchange

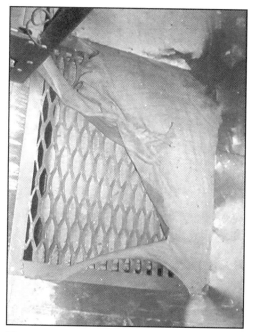

Fig. 3–12. Covered hoistway vent exchange opening

Some of the other variations of elevators, whether they are basement traction machines, the basement drum, overhead traction, or others, are too many to list (fig. 3–13). This book, after all, is aimed at firefighters, and for basement traction schematics, we point you to other publications (see References) that will be technically correct in their approach (fig. 3–14). We want to teach you safety, not how to be an elevator mechanic.

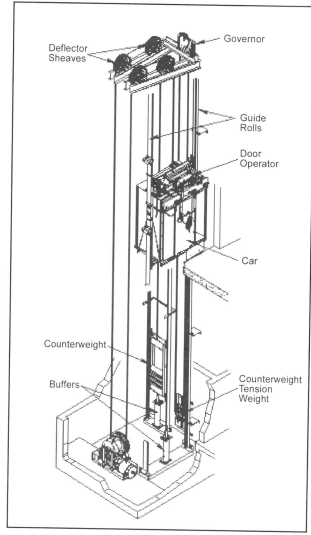

Fig. 3–14. Courtesy of KONE Corp.

Fig. 3–13. Basement machine (geared) (courtesy of GAL Manufacturing Corp.)

Summary

The traction machine room has gone from a place of large machines and temperature extremes to a very quiet room, with air conditioning and heat with temperature control.

The basic types of machines are

- Geared, traveling typically up to 450 ft/min
- Gearless, traveling typically at speeds of 450 ft/min and higher

The machine room is usually located directly above the hoistway, or in the case of MRLs, the actual machine is in the hoistway. Many now have their small motors in places no one could have imagined 10 years ago.

Always carry your forcible entry tools with you, because no locked machine room door should prevent a fire department from performing lockout/tagout on the power source. That power source is the mainline disconnect, located in the machine room or space or closet, whatever the configuration might be. Older disconnects can present a challenge due to the construction of the electrical box not allowing lockout/tagout, requiring a physical presence during the operation.

When entering machine rooms, be very aware that there may be up to 600 volts AC, all of it waiting to marry itself to the tool you are carrying. Use the tools to get in, then leave them outside the machine room door.

Finally, we never move or lower an elevator. That is the job of the trained elevator mechanic alone.

Review Questions

1. List the Big 8, and explain their use.

2. Describe the operation of a shunt trip device.

3. What is the AC limit in an elevator hoistway?

4. A gearless elevator runs at what speed or higher?

5. List the dangers of a hoistway vent exchange opening.

Field Exercise

Within your response district, locate and document the following equipment:

- Geared machine
- Gearless Machine
- Basement machine
- MRL

Endnotes

[1] McCain, Z. *Elevators 101.* Elevator World: 2004, p. 75.

Chapter 4
Types of Doors, Interlocks, and Restrictors

Fig. 4–1. Two-speed center-opening slide door set

The hoistway doors are the first part of the elevator system that greets the firefighter on arrival at a response for a stalled elevator. The elevator system uses a variety of hoistway and car door designs to accomplish the desired effect of protecting the passengers. This is not a complicated process for firefighters to assess, because there are only a basic number of door designs. The following list includes the types that you may encounter.

- Swing door—can be power or manual
- Single-speed side-slide
- Two-speed side-slide
- Center-opening
- Two-speed slide center-opening (fig. 4–1)
- Biparting freight
- The other creatures that are still out there

Swing Doors

The swing door has been around the elevator world for many years, and it has been the scene of many deaths and injuries, mainly involving children (fig. 4–2). They are found in many types of installations, both large and small, such as housing projects, private homes, small inns, or other occupancies. They are used in conjunction with an elevator car door or a sliding gate of one variety or another and a swing door, which when closed completes the system. The main problem with this system is the amount of space in the floor threshold area between the back of the closed hoistway door and the front of the car gate or door. The latest ASME edition limits this space to three inches maximum, where the pre-1950s era code allowed an indeterminate amount of space. (See chapter 10, on residential elevators.)

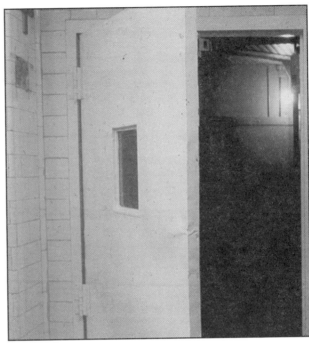

Fig. 4–2. Swing door elevator

Single-Speed Side-Slide

The single-speed side-slide door is probably the most common elevator door used in the elevator business today (fig. 4–3). We see them constantly in low-rise office buildings, and we also see them as the one elevator in older apartments and housing developments. The hoistway door slides in whatever direction that the system is designed for, and the car door is of the same design.

Fig. 4–3. Single-speed side-slide

Two-Speed Side-Slide

Two-speed side-slide is a common door in many installations around the country (fig. 4–4). Take note of the insert panel (called the "fast panel"), which moves at twice the speed of the front panel (called the "slow panel") initially, then the two speeds equalize as the pair clear the doorway opening. The speeds are reversed as the doors close. Note the high-speed panel via mechanics fingers.

Fig. 4–4. Two-speed slide doors

Center-Opening

Center-opening slide doors open from the center of the door system, sliding in opposite directions. They provide a clean, wide opening for passengers to enter or exit (fig. 4–5).

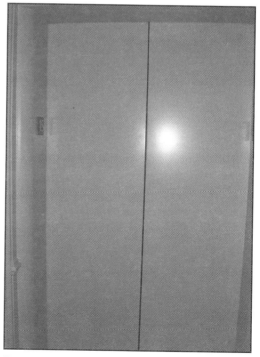

Fig. 4–5. Center-opening door

Two-Speed Slide Center-Opening

The two-speed slide center-opening door is a combination of the center-opening and the two-speed side-slide (fig. 4–6). It is usually provided when there is insufficient hoistway width for door panels when fully open. If the situation warrants forcible entry, make sure that all tools are used only at the top jamb (closing point) end of the door, not at the bottom or sides of the doors. This will only complicate the problem and will not result in getting the occupants out of the car.

Fig. 4–6. Two-speed slide center-opening door

Biparting Freight

Biparting doors are commonly found in what are generally referred to as freight elevators (fig. 4–7). When powered, the car and hoistway doors are powered independently. The manual variety work in tandem with each other. They are counterbalanced, and the downward movement of one panel transfers through a system of chains and pulleys into the upward movement of the other. We always must be mindful of the elevator that no one has seen in years and is just waiting to injure or kill someone who happens to use it.

Fig. 4–7. Biparting vertically sliding freight doors

The Other Creatures Still Out There

On the other side of the street-level door is the unprotected hoistway (fig. 4–8).

Fig. 4–8. Street-level hoistway entrance

Elevator Doorway Protection

Whatever door design is used for the hoistway door, the car door will usually be of the same type. This is necessary, as they are operated together by the clutch that sits on the hoistway side face of the car door and interacts with the interlock located on the back of the hoistway door. As the elevator moves up and down its hoistway answering calls, it meshes together, via the clutch, at each floor, opening and closing the doors as a unit (figs. 4–9 and 4–10). This is done by means of our friend, the car door operator, which sits on top of the elevator car. The photoelectric eye (P.E. Eye) unit shown in figure 4–11 provided early protection of the car-door opening for the passenger. Other varieties include "nudging" door protection, where the door operator actually keeps nudging the obstruction until it has a free movement, or it shuts down after a timer has operated. Nudging is actually "reduced force closing" (2.2 foot pounds of kinetic energy [KE]) and is required by code anytime you close doors without door protection. The modern type of door protection is

the light-emitting diode (LED) shown in figure 4–12. This unit actually provides a head-to-toe curtain of protection to the doorway, preventing the doors from closing if an object of any size breaks the light flow of the unit (fig. 4–13).

Fig. 4–9. Vane clutch

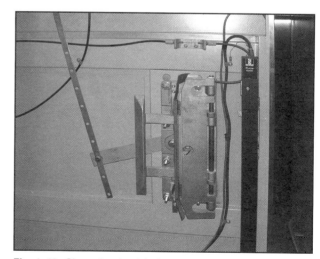

Fig. 4–10. Shoe clutch with door protection

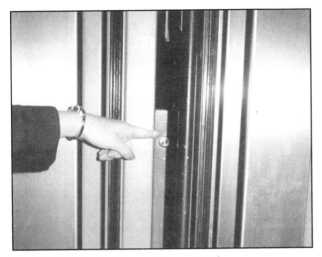

Fig. 4–11. Photoelectric eye door protection

Fig. 4–13. Electric door protection mirror

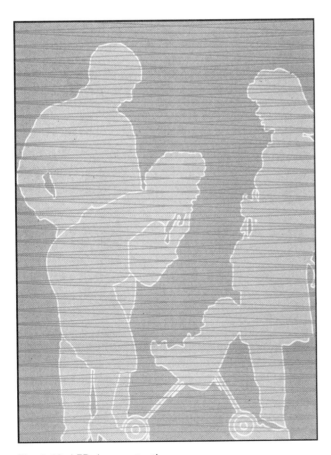

Fig. 4–12. LED door protection

The linear door closer and its restrictor in the photo in figure 4–14 show us that they are becoming smaller and more sophisticated with each generation of door closers.

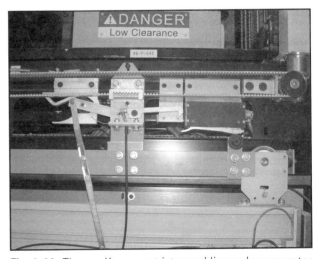

Fig. 4–14. ThyssenKrupp restrictor and linear door operator

The door closer is the tracked belt assembly, and the restrictor keeper is at the top, center-right, just below the door operator belt. This type of door operator is used for light-duty work, while others are larger, intended for an anticipated heavier workload (fig. 4–15).

Fig. 4–15. Car top inspection station

The car door operator is shown on top of the elevator in photo. Also shown is the car top inspection station, located on the crosshead beam.

Fig. 4–17. Shoe clutch with restrictor arm

Fig. 4–16. Shoe clutch

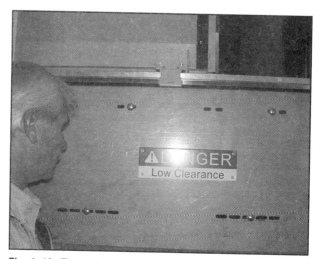

Fig. 4–18. ThyssenKrupp toe guard

A center-opening door configuration is shown in the photo in figure 4–16. Note the shoe clutch, which will engage the landing door pickup roller assembly as the car levels into a landing. Note the restrictor in place on the clutch with its extension arm leading up to the actual restrictor on the door-closer track (fig. 4–17). Shown in figure 4–18 is the car apron or toe guard, which is a physical barrier that hangs down a maximum of 48 inches (ASME A17.1–2000) from the front of the bottom of the car assembly. Most are 21 inches long. If the car were up in the opening, it would create a barrier to the open hoistway from the public. From the hoistway side, the hoistway door has a definite working appearance, unlike the smooth, polished lobby side of the same door. The doors are hung on a horizontal door hanger, as shown in the photo in figure 4–19. The keeper is fastened onto the back of the hoistway door at the leading edge of the panel. As the door closes, it drives the keeper of the lock into the interlock that is mounted onto the hoistway header or track, thus confirming that the door is both closed and locked.

All door panels on modern passenger elevators hang on rollers, which roll on tracks mounted at the top of the car or hoistway opening. At the bottom of the door panels, they are guided in the sill slot by door guides. Door guides, which are often referred to as "gibs," normally have a plastic or metal member attached to the door panel that moves in the sill slot. During a fire, the plastic component of the gibs may melt out, causing a lack of stability at the bottom of the door panel. To pass the fire door test, there is a requirement for a secondary safety retainer on the bottom of the door to ensure that the door cannot swing open from the bottom. ASME A17.1 requires the door have a spring closer that will automatically close the hoistway doors if they become separated from the car doors.

Fig. 4-19. Photo of door system of GAL elevator

In Fig. 4–19, you can clearly see the system as it is constructed. The door-return system, as shown in the figure 4–20, consists of the door itself, the interlock and keeper assembly, interlock release assembly and release rollers, track mounting plate and track, along with the sheave and reel closer (the reel closer replaces the spring closer and does the same job). It also includes the hoistway door unlocking device keyhole (ASME A17.1, section 2.12.6.1) assembly, which is required on all floors for emergency access unless restricted by the authority having jurisdiction (AHJ). Some AHJs (Massachusetts) have only recently approved their use in new installations, but not for existing ones. Consult your local AHJ for its position on this issue.

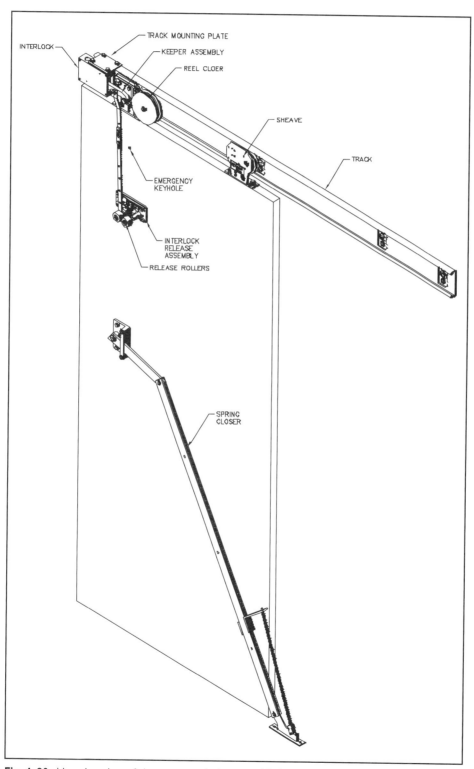

Fig. 4–20. Line drawing of door system hoistway side (courtesy of GAL Manufacturing)

Interlocks and Restrictors

It would be a disservice to all if we did not go over interlocks and restrictors in greater detail, as both play an ever-increasing role in our response to elevator emergencies. As touched on earlier in this chapter, the keeper is located on the back of each hoistway door in the elevator system (fig. 4–21). If by chance the elevator concerned has two openings (front and rear), then there will be a keeper and interlock for each.

In the American elevator industry, it was always the golden rule with interlocks that the release roller was the one closest to the final closure doorjamb. These rollers, commonly called pickup rollers, are engaged by the clutch on the face of the car door as the elevator glides into a landing. With others now in the marketplace, they may use the release roller furthest from the final closure doorjamb. It really does not affect your operations that greatly, as most release rollers are on an extension arm coming down the door from the keeper. The interlock itself is usually hidden behind a dust cover for protection from the hoistway elements (figs. 4–22, 4–23, and 4–24). When in doubt, just *pull* or *push* (lift usually) the extension arm one way or the other, hold that position, and slide the door open.

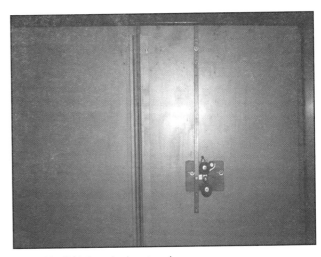

Fig.4–22. GAL interlock extension arm

Fig. 4–21. Interlock with extension arm

From the fire service's standpoint, this is not a complicated process, because there are only a few major manufacturers left, and they either use their own interlock or a component supplier's equipment to meet their needs for a specific installation. It is not unusual to see two different construction sites from the same elevator company, one where they are using their interlock and the other where they use a supplier's model.

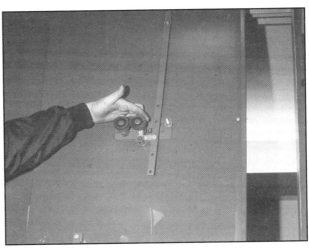

Fig. 4–23. Fujitec-America interlock

Fig. 4–24. Otis 6940 interlock with dust cover and electric box cover removed

Firefighters need to look closely at figure 4–24 to understand the role that the interlock plays in the elevator system. The basic construction consists of the keeper, safety circuit contacts, electrical box, and release (pickup) rollers. Notice how firmly this Otis lock is mounted into the overhead of the doorjamb with large bolts screwed up into the cement.

Fig. 4–25. Otis with dust cover

Interlocks

In figure 4–25, the dust cover is in place. The interlock performs two functions at the same time:

- First, it keeps the hoistway door closed by a mechanical means called the *keeper*. It has a lip that catches physically onto the lip of the interlock box. Thus, the mechanical part of the interlock has been satisfied. If a person were to bump against the door, the keeper would keep the door closed (fig. 4–26).

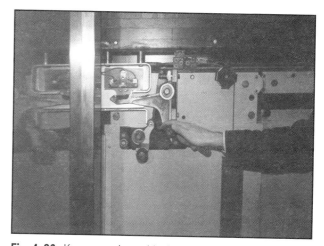

Fig. 4–26. Keeper activated by hand

- Second, it is a part of the safety circuit that monitors all openings into the hoistway. The keeper has shorting contacts that must be in contact with corresponding contacts which are a part of the interlock box. When the keepers contact "makes" with the interlock contacts, the safety circuit is closed, and the car is allowed to run. If someone in a 20-story building forced the floor door open at the first floor, the car which happens to be at the 18th floor will perform an emergency stop. In the machine room, the safety circuit power keeps the brake shoe, which is located on the hoisting machine, physically off the brake. When the safety circuit is opened, as when someone forces open a hoistway door or car door, it applies the brake.

Fig. 4–27. Schindler (Westinghouse) drive blocks

The example shown in figure 4–27 is a set of drive blocks that were used with Westinghouse Elevator Company equipment. Westinghouse was bought by Schindler Elevator Company, and you will still find drive blocks out in the field today. As for us, the fire service, it usually comes down to the following:

- Vane clutches work with drive blocks.

- Vane clutches work with release rollers.

- Shoe clutches work with release rollers.

- Shoe clutches *do not* work with drive blocks.

In chapter 8, Freight Elevators, we will introduce you to the interlock used with freights and dumbwaiters, which uses the retiring arm and cam.

Restrictors

The elevator industry has been placing restrictors on elevators since the publication of the 1980 edition of A17.1 Code. The reason why they are there is to keep people from getting out of a stalled or malfunctioning elevator that is out of the unlocking zone (3–18 inches above or below a landing) and putting themselves into great danger. There have been many instances of the riding public, particularly students and other younger people, who have fallen under the opening below the car down the hoistway to their deaths. By restricting their ability to roll the car doors open, they are kept safely inside the car. (See figs. 4–28 though 4–31.)

- If a car is *within* the unlocking zone, the restrictor is not engaged, and the car door, if rolled properly, will open. This is allowed, because there is not room for them to fall under the car and down the hoistway at this point.

- However, if the car is *out* of the unlocking zone, the restrictor will engage the car door as the occupant attempts to open the doors, and stop it with only 4 inches of opening allowed. The occupant will not be able to reach the restrictor to disable it or compromise its function.

- According to ASME-A17.1–2004, section 2.12.5, the restrictor must be capable of being easily operated by rescuers or mechanics from the landing side of the hoistway after they have shut off the power to the affected elevator and opened the hoistway door (fig. 4–28).

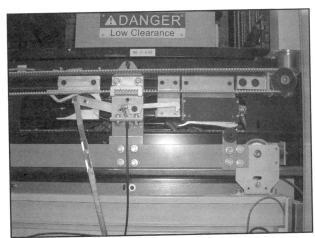

Fig. 4–28. The ThyseenKrupp restrictor was found on a new MRL installation. It would be easily manipulated from the landing sill area by lifting the keeper lip on the right, top center.

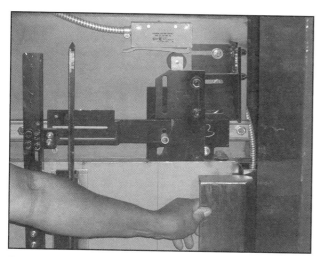

Fig. 4–29. An Otis mechanical restrictor. The mechanical restrictor can be opened by lifting the round clear plastic lip up and out of the slot. While doing this, push door to the left to open.

Fig. 4–30. The Otis electronic restrictor may be opened by pushing the black plastic tab that sticks down below the unit, which actually blocks the door from rolling open when it hits the restrictor bar.

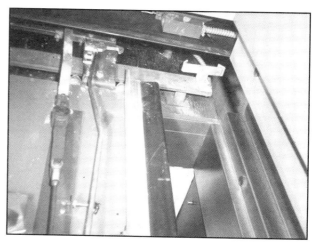

Fig. 4–31. A basic restrictor. When the early restrictors were put onto existing equipment, they were of a very basic design, as shown. Lift keeper to disengage.

In the two images in figures 4–32 and 4–33, two phases of restrictor activity are shown. In figure 4–32, it shows that if one were to roll the car door open, the restrictor neck would clear the piece of steel welded onto the rail. This simulation is within the unlocking zone, and the car door would open all the way. In figure 4–33, (courtesy of GAL Manufacturing Corp.) it is clear to the eye that if the door were opened any further, the metal and the restrictor neck would hit, and the door would be restricted to four inches of movement and stopped. This simulation is outside the unlocking zone, and the occupants would be prevented from exiting the car into danger.

An important issue regarding restrictors is that if firefighters are on Phase II firefighters' service operations, and their elevator stops for whatever reason outside of the unlocking zone, the restrictor is active. It will require forcible exit work to get the car door open. There will be a problem regarding getting a bite, or a "purchase" as some say, due to the door moving the four inches open before stopping. Toronto Fire Service firefighters tried to force a restrictor during a drill and found it very difficult. Solution: Make sure you have brought your small hydraulic ram tool as part of your forcible entry equipment. You may need it to get out of the elevator!

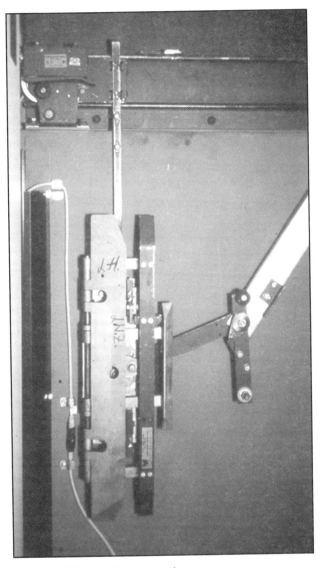

Fig 4–32. GAL restrictor not active

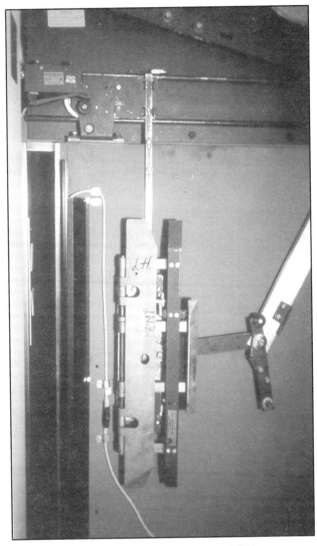

Fig. 4–33. GAL restrictor active (courtesy of GAL Manufacturing Corp.)

The issue of changing the action of restrictors when on Phase II operations came before the ASME A17.1-Emergency Operations Committee (EOC) during the 2005 meeting year. A report was filed by the fire service members on the EOC, but the motion was defeated because of a lack of any documented incidents to back up our concerns. The ASME A17.1 EOC felt that door restrictors were one of the most significant improvements that have saved countless lives and that change should not be made in the requirement for the restrictor. This issue will probably come back before the board at another time, and we can hope for changes possibly at that time.

Summary

As firefighters, the arena of doors, interlocks, and restrictors is as important an area for us to review periodically as anything else in this book. We urge you not to stop with this manual, but seek out the new information as it appears in the field. This information is critical to our safe operation around elevators.

Remember, that there are six basic door designs:

- Swing door—can be power or manual

- Single-speed side-slide

- Two-speed side-slide

- Center-opening

- Two-speed slide center-opening

- Biparting freight

Knowing each type and how each differs from the other will help you work through your incident and determine what type of interlock you may be facing. When examining a door for interlock location, find the leading edge of the door, where final contact is made by the closed door. This will usually be the location of the interlock, at the top corner of the panel. The extension arm located on the surface of the back of the hoistway door will open the interlock. After activation, it must be held in the open position while the hoistway door is slid open. Remember that the unlocking zone is 3 to 18 inches long. If a car is 19 inches off the floor, then the restrictor, if provided will be active. ASME A17.1 mandates that all restrictors must be operable from the landing once you have opened the hoistway door. They come in a variety of shapes but basically are of three designs:

- Mechanical

- Electronic

- Collapsible

Whatever the design, they are there to save the lives of the riding public who might otherwise endanger themselves by trying to exit the car. Unfortunately, this device creates a potential problem for us when on Phase II operation.

Review Questions

1. Describe the function of the safety circuit as it relates to a floor door.

2. List the six hoistway doors discussed in this chapter.

3. Name the types of restrictors that were reviewed.

4. What interlock uses the drive block instead of the pickup roller?

5. What is the purpose of the restrictor on the car door?

Field Exercise

Working with local members of the elevator community, establish a database or picture base of the various doors, interlocks, and restrictors that you encounter. Part of this exercise may be accomplished over the Internet, while actual field sites should make up 50% of the projected time spent.

Chapter 5
Hallway and Elevator Lobby Features

The elevator lobby contains many important parts of the elevator system that we are examining (fig. 5–1). The means used to summon the elevator by the waiting passenger—the floor call buttons—are located here on the wall adjacent to or between the car(s). At this location, you may also find the emergency power selector, if provided. Also in this lobby will be access to any fire command center if the building is large enough to warrant one, the control desk for any building security systems, and the position indicators, directional or arrival lanterns, and firefighters' emergency operation Phase I capture station. (Use of elevators during a building fire will be covered in chapter 24.)

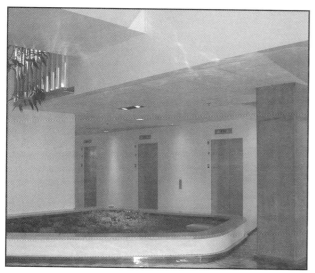

Fig. 5–1. The elevator lobby

Fire Command Center

As we arrive at this building for a reported fire, we report to the building fire command center (FCC), to check the fire alarm annunciator panel for the fire alarm initiating device (FAID) on the floor of activation (fig. 5–2). (Get used to it—FAID is the current term.) The FCC will be located, ideally through preplanning, convenient to our access into the building. Unfortunately, we have all hiked long distances to find it, gather our information, and then hike back to the elevator lobby, sometimes from subcellar locations to start operations. Of even more concern is that these locations were usually agreed to by the local authority having jurisdiction (AHJ), our fire department. At the FCC, we should find the following components to assist us in fighting a fire or locating a stalled elevator:

- Elevator position panel (fig. 5–3)

- Fire alarm annunciator panel (fig. 5–4)

- Emergency power selector panel (See A17.1, 2.27.2.4.1)

- Remote firefighters' emergency operation Phase I switch(es) (fig. 5–5) (See A17.1, 2.27.3.1.2)

- Emergency communications panel

All of these and other features will help the incident commander (IC) run the incident and keep track of who is where, what they are doing, and what the conditions are on the floors above. An important point to make is that we check the annunciator ourselves and listen to, but *never* take the word of, building security.

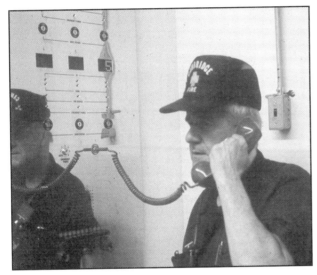

Fig. 5–2. Fire command center

Fig. 5–3. Elevator position, 3502 remote and power panel selector

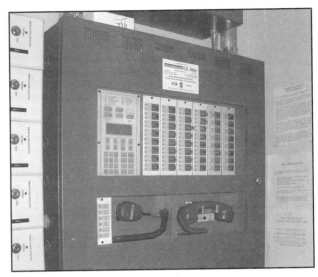

Fig. 5–4. FAID annunciator panel

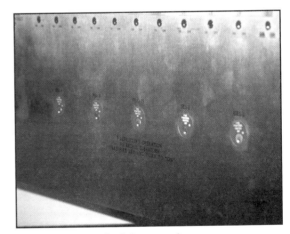

Fig. 5–5 Remote firefighter service

If we have been summoned there for an elevator emergency, many of those same features may assist us in our rescue or removal operation.

At this point we should examine the types of buttons used by the public to summon the car to the lobby. The listing is simple—there are only two types used in the elevator world:

Fig. 5–6. Mechanical button

Fig. 5–7 Otis electronic touch button.

- The mechanical button requires pressure to be placed on the button, which causes a movement inward, thus completing a bimetallic contact to institute a car or floor call (fig. 5–6).

- The electronic button was originally developed by Otis Elevator. This button does not move when touched but uses the individuals' static electric charge to call the elevator (fig. 5–7). This is done by storing the needed charge on the button surface, but leaving a slight gap in the charge, which is completed by the person. In the 1970s, these buttons were the source of confusion and misinformation throughout the fire service. A misconception existed that they operated from the heat on one's finger, causing the car call for the elevator. Today, we know the correct story, and hopefully we can put that incorrect thought to bed.

Elevators are numbered according to the A17.1 Safety Code for Elevators, usually from the left, clockwise. They must be marked accordingly. The important thing to note is this: Where is the accepted entrance way into the lobby? What was the starting point for the numbering of the elevators? There are diagramed examples of car-numbering patterns, but unfortunately they are not universally followed (figs. 5–8, 5–9, and 5–10). Check with your AHJ for local practices.

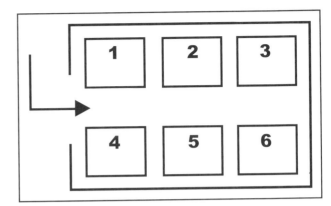

Fig. 5–10. Six-car bank

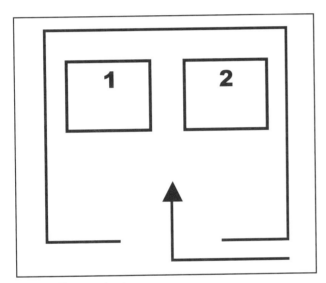

Fig.5–8. Two-car bank

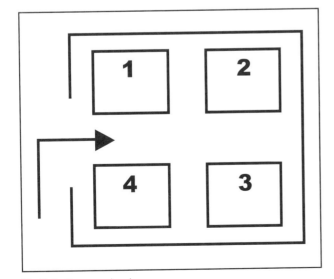

Fig.5–9. Four-car bank

Fig. 5–11. Miconic 10 hallway keypad

Press a desired floor number, and a letter will appear designating which elevator to take (e.g., A, B, or C). We may see what appears to be newer concepts, but they all work either electronically (card identity types) or mechanically (Schindler Miconic 10; see chapter 9, New Technology) (fig. 5–11). At the other floors throughout the building, the corresponding floor call buttons will be of the same type in the cars and on all floors as was found in the lobby. The common denominator of all of these buttons is that they do not do well in the fire environment. They will melt from high heat temperatures, causing them to deform and collapse, with the result that a false call is entered for the fire floor. In the past, prior to the inception of automatic recall in the code, there were terrible incidents of elevators being called to the fire floor, opening their doors to a fire or smoke-involved elevator lobby, with the resulting deaths and injuries to those unfortunate enough to be in the car.

Activation of the smoke detector (FAID) located in the elevator lobby now results in the immediate Phase I recall of all of the cars in that bank or group, so the public is no longer in danger during the incipient fire. While you are still in the station, donning your turnout gear, they are already safe in the designated level lobby of the building, and the fate that many innocent people suffered in the past has been eliminated (fig. 5–12).

Fig. 5–12. Clearwater, Florida, fifth-floor lobby (courtesy of U.S. Fire Administration (USFA)[1] and TriData Corp.)

Remember: In older buildings that were constructed prior to automatic recall and firefighters' emergency operation, the danger still exists.

Fig. 5–13. Hall position indicator (PI)

There are a number of other features that you may find in an elevator lobby. There may be a hall position indicator, commonly referred to as the hall PI (fig. 5–13). We say "may be," because like purchasing an automobile, some things in an elevator system come as standard, while others are accessories and cost extra. In place of a PI, you may have a directional or arrival lantern that is used to notify the public of an approaching car answering their request (fig. 5–14).

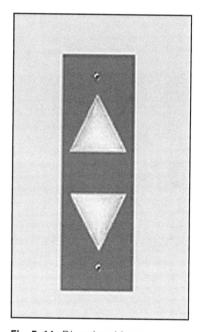

Fig. 5–14. Directional lantern

The most important development came about when automatic recall was adopted, which removed the cars and their occupants from the dangers of the fire environment at the first wisp of smoke that hit a lobby dedicated detector (ANSI A17.1a-1977, sect. 2.27.3.) This has undoubtedly been the most important development in safeguarding the lives of the public (fig. 5–15).

Fig. 5–15. Elevator lobby dedicated smoke detector (FAID)

Fig. 5–16. Fatal fire floor elevator lobby (photo by Ed Fowler–CFD)

In figure 5–16 a lobby arson fire claimed the life of a young mother going out to work. A 22-story Housing and Urban Development (HUD) building, it was built without sprinklers in the late 1960s. There are thousands of these buildings across the United States today, waiting to claim other victims. How many hundreds of lives were lost from occupied cars stopping at landings of fire floors, in the early days of recording fires in buildings and their effects on elevators? In succeeding years (ASME A17.1–2000), the code has been adjusted to eliminate the short circuiting of lobby equipment from controlling the operation of Phase II operation, which is particularly important to the fire service community. In the past, hallway fixtures were the source of many

problems when they would fall victim to the elements associated with fire. To overcome this problem, various corrections were developed to safeguard the riding public and the members of the fire service who might be in these cars during a fire.

The access switch key is not intended for use by anyone other than an elevator mechanic (fig. 5–17). It is used when the mechanic wants either to ride on top of the elevator to perform maintenance and inspection duties or view the underside of the elevator. The access switches are located at the first two landings to gain access to the pit and top of the car.

Fig. 5–17. Mechanic's access switch

This is a restricted key. It does not accept the Phase I key, so please do not try it to use it with your key. If done, you will end up with no key, and the jammed key receptacle will be unavailable to the mechanic. An involved process then follows that will allow the mechanic access to the top of the car or pit of the elevator. The car may no longer answer floor calls from the lobbies or any firefighters' emergency operation alarms or manual activations. A bell on top of the elevator will ring a signal when Phase I is activated that the mechanic knows means Phase I of firefighters' emergency operation is being activated in the building. He will then bring the car down to the designated level (lobby) and return it to normal service for the firefighters to use under Phase II. It cannot be emphasized more that this feature is for trained elevator mechanics use only.

Fig. 5–18. Emergency medical service (EMS) key and marker

The EMT or EMS key, if provided, is located in the lobby of a building, in the same area as firefighters' emergency operation (fig. 5–18). This will be located to control one elevator in the bank, and it will follow a specific set of guidelines similar to hospital/emergency service (see Glossary). When turned on, it has priority over firefighters' emergency operation, unless Phase I is already turned on. It becomes a case of whoever gets there first gets the elevator.

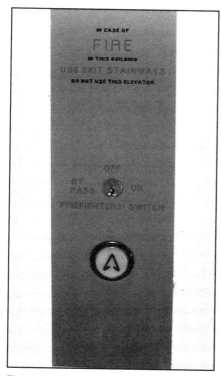

Fig. 5–19. Firefighters' emergency operations—Phase I

Firefighters' emergency operations Phase I is located in the lobby at the designated level (DL). This feature provides the fire service with a means to *manually* take control over all of the elevators in that particular bank or group of cars. The station shown in figure 5–19 is what the firefighter will find on all installations except the newest ones (2005). They will appear as in figure 5–20, with the feature known as *by-pass* replaced with the *reset* feature. A further explanation will be provided in chapter 24, Use of Elevators During a Building Fire.

Fig. 5–20. Firefighters' emergency operations—Phase I with reset position

The changes to firefighters' emergency operations over the past few years have been significant. The function of recall remains the same, but the appearance of the Phase I recall station has changed. The wording change from *by-pass* to *reset* was to take the A17.1 Emergency Operations Committee out of the smoke detector business, because it had no experience with the many changes that were occurring daily with those devices. It was felt that the placement and other parameters of smoke detector operations would be better served by referring to NFPA 72, the recognized national standard in that field.

Reset is used to reset the elevator systems only. If your fire alarm panel is not clear, you will not be able to reset the elevators back to automatic operation until the fire panel has been programmed to delete the defective or damaged smoke or heat detector, or it has been replaced.

Summary

It is vital that we know our buildings as well as we know the station that we work in. Those who work in the city setting will find that more difficult to accomplish than those who work in a less intense suburban-like setting. We have often stood through countless delays in the lobby as officers sought out the location of the fire command post or the sprinkler room, so critical to us for the information that it contains.

The advent of automatic recall in ANSI-A17.1a-1977 was a huge leap forward for the safety of the riding public in elevators. We must remember that this advance *did not* apply to existing elevators, and as we have mentioned a number of times, A17.3-Safety Code for Existing Elevators has not been adopted by more than a few AHJs. This means that you will still run into buildings where the elevators may be drawn to the fire floor by the fire conditions damaging floor call buttons.

The buttons comprise two main types, mechanical and electronic, and even with the new looks (Miconic 10, zip cards), they are still one type or the other.

The public may still press the fire floor and arrive there to face the smoke and heat of a fire lobby. Past history will repeat itself, unfortunately, and deaths and injuries will result.

The location of the fire command post with the FAID annunciator, and the other related equipment for the elevators such as emergency power selector, elevator location panel, water flow alarms, and the like, will be of no use to you if you are not familiar with what the indications mean and know how to read them.

It is important to know the difference among the Phase I key, the EMS, and hospital service key switches. All are meant for emergency response, but they are different varieties.

Be aware of the difference and the function of a Phase I key switch and the access switch. One we live by, the other we do not touch.

Finally, and of great importance, is to know the number of the car that you are getting into or looking for. It could mean your life!

Review Questions

1. What is FAID?

2. Describe the various types of floor call buttons and how they function.

3. What has been the most important development in safeguarding the public?

4. Define the automatic recall feature.

5. Explain the difference between the by-pass and reset markings on a Phase I recall station.

Field Exercise

Visit three buildings in your first-alarm district and note the differences between the elevator systems, if any. Mark down their machine room locations and lobby features.

Endnotes

[1] "Multiple Fatality High-Rise Condominium Fire." Clearwater, Florida. USFA-TR-148/June 2002.

Chapter 6
Car Interior Components and Finish

Car Operating Panel

The elevator car or cab interior has many familiar features that we should review. The most common piece of equipment that we as firefighters see and use every day is the car operating panel (COP; fig. 6–1). The car operating panel, sometimes referred to as the gang station, is where passengers place their requests for a floor call. It is where all the necessary functions are located, so that a passenger, firefighter, or elevator mechanic can find needed aspects of the panel. Starting at the top is the emergency light, which is powered for up to four hours by a battery pack inside the panel. It also may be located in the ceiling of the car, but it must illuminate the COP. The next feature is the firefighters' emergency operation component, which will be covered in greater detail in chapter 24. There may also be a key switch for EMT or hospital service, inspection, or attendant operation.

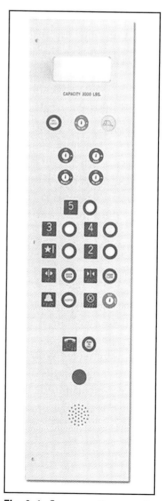

Fig. 6–1. Car operating panel

Next are a series of key switches that are used in the service life of the car by the mechanics, followed by a number of floor call buttons, door open/close buttons, alarm bell, and the in-car keyed stop switch for a mechanic, which when activated will cause power to be removed from the elevator driving machine motor and brake (ASME A17.1, sect. 2.26.2), bringing the elevator to a stop. Older panels may have an emergency stop button, which has been removed from public access under ASME A17.1–1987 and later editions. At the very bottom will be a push for help button and speaker port for two-way communication between the passengers and those responding to their needs. Note that the ASME A17.1a-2002 Code and later editions requires that the *building owner* be responsible for providing a 24-hour means of responding to any call using such a system. This is usually handled by an answering service of a service company or a private answering service that must answer within 30 seconds and who can take the appropriate actions. The communications system is required to be provided with a 4-hour battery-powered supply for the emergency communications system if it operates on normal power and the audible signaling system for 1-hour (ASME A17.1–2000 and later editions). It is not recommended that a fire department provide this point of contact, as it is the building owner's responsibility.

Some elevators may have a small door at the bottom of the COP, which when opened with a key reveals an independent service panel (fig.6–2). By operating a toggle or key switch, the elevator can be moved from automatic to attendant operation. This switch is used many times by building management when they are moving people in or out of the building, or for other maintenance-related activities. It provides them with an uninterrupted measure of control of that specific elevator for whatever purpose they need. The fire service should *never* use this feature when entering a building on a fire response, given that the elevator will have *none* of the protection that firefighters' emergency operation Phase II offers them. This use, coupled with a well-intentioned building staff, has led to the deaths of all in the elevator when the custodian pressed the fire floor number, and the doors opened at the fire floor, rather than *two floors* below, as we do.

Fig.6–2. Independent service panel

Emergency medical service/hospital service—also known as "code blue" in some circles, allows trained medical staff in medical facilities to take control of an elevator that has this feature. In the selected bank(s) of elevator(s), an access switch marked "Medical Emergency" will allow trained personnel with the key to call the elevator to that floor (fig.6–3). This would then allow the staff to respond to an emergency elsewhere in the bank or take a patient directly to urgent care.

Fig.6–3. Medical Emergency

Firefighters' Emergency Operation Phase II

Fig. 6–4. Fire emergency operation Phase II panel

The Phase II fire operations panel has also undergone some changes over the past few years. It is important that all firefighters become aware of the different appearance that elevators will present to them. The Phase II station that we are all familiar with is shown in figure 6–4. This Phase II station has all of the components that we are all familiar with, including

- Three-position Phase II key switch (off-door hold-on)

- Call cancel button

- Fire hat symbol (is it flashing?)

- Operating directions

- Emergency light

Other components, such as door-open, door-close, and "push to talk" emergency buttons, are usually located at the bottom of the COP. This spread-out location of components was never intentional, but just ended up that way from normal location of the components for nonfirefighter use. Each elevator and components supplier had its own reason for the button placement at the time, but on another version of its panels, it may be placed elsewhere. This created a hodge-podge of different panels that the firefighters were continually trying to become familiar with.

Phase II fire operation panel—with the advent of the A17.1–2004 Safety Code for Elevators and Escalators, a long-awaited change took place. For five years, members of the A17.1 Emergency Operations Committee worked with industry and the fire service to come up with a better way of placing the Phase II components that firefighters needed to complete their operations. One of the many complaints we all had was that components were all over the map on the face of the COP. With the advent of the firefighters' operations panel, that problem has been resolved (figs. 6–5 and 6–6). All the Phase II components are located in one panel, along with an emergency stop switch. They are located behind a locked cover that is opened by the same key that operates Phase I. This panel will be covered in more detail in chapter 24, Use of Elevators During a Building Fire.

Figures 6–5 and 6–6 show examples of a fire operations panel by two of the manufacturers, GAL Manufacturing Corporation, and C.J. Anderson & Company.

The interior look of the car will depend on how much money will be spent on the elevator, and this will be dictated by the building owner and architect. If it is a corporate headquarters of a large firm, it may be very plush, with teak wood walls, stock market LED screen reports, and indirect lighting with soft music to please the riders. The size of the car itself is determined by the number of people to be moved or if required to carry an ambulance gurney.

Passenger elevator cars that have been constructed in accordance with the ASME A17.1 Safety Code for Elevators and Escalators must meet stringent combustibility requirements (fig. 6–7). Materials exposed to the interior of the car and the hoistway, in their end-use composition, are limited to a flame spread rating of 0 to 75 and smoke development rating of 0 to 450. Some jurisdictions have mandated that the area be large enough to fit an EMT ambulance gurney or stretcher (24 inches wide by 84 inches long) in its horizontal position (EMT service or hospital service). This will enable EMS operations to perform CPR in the proper supine position.

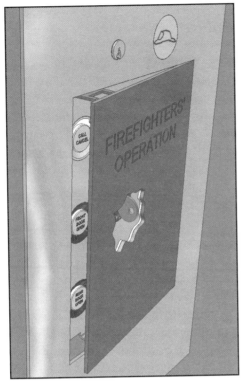

Fig.6–5. Fire operation panel (courtesy of GAL Manufacturing Corp.)

Fig. 6–6. Fire operations panel (open) (courtesy of C.J. Anderson & Company)

Fig 6–7. Car interior

Fig. 6–8. Firefighter pushing light baffle aside

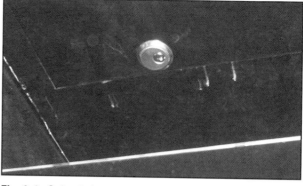

Fig. 6–9. Seismic key access to top of car hatch

However, we have all been in a minimum-expense housing elevator, smaller than a shoe box, with cheap wall construction, poor lighting design, distinct odors, no ventilation, no music, and so forth. The door facing us will be the inside face of the car door, and it usually will be of the same design as the face of the door that we saw as we entered the car. The ceiling may have false ceiling panels, which, if pushed up, will reveal the emergency exit cover (fig. 6–8). This will be normally positioned in the rear section of the ceiling, and will be locked from the *outside,* to prevent accidents from happening. Some jurisdictions have mandated that the Phase II key operate a key tumbler in that hatch, to provide firefighters a way out in an emergency (fig. 6–9). In the past, there was a side emergency exit panel, but these are no longer installed due to the number of people killed when using these exits improperly. Some authorities having jurisdiction (AHJs) now call for existing side emergency exits to be bolted or sealed from the outside of the car to prevent their use. When preplanning your buildings, make note of the ones with extremely high ceilings that would require a small folding ladder to get out through the car top emergency exit. As difficult as it may sound to bring the folding or pencil ladder with you, it is even more difficult trying to climb on each others shoulders to get out. The subject of using top emergency exits and the hazards of such operations will be covered in its own section (chapter 15).

Summary

Although we are looking at a relatively small space, or box, the car contains features that are very important to our safety in an elevator. The first thing that we see is the car operating panel, or COP as it is commonly called. Another name used is the gang station. On it are located all of the features necessary to those who ride or work in an elevator.

The panel will have the following:

- Floor selection buttons

- Phase II firefighters' emergency operation features

- Various special function key switches

- Emergency communications panel

- Emergency lighting unit

The panel is located on the wall of the car adjacent to the entrance to the car. The construction of the car will be in adherence with the ASME A17.1 Code. There may be a digital screen display that is used to advertise the local attractions, as well as giving the latest stock reports and sporting event scores.

Finally, in the roof of the car there will be an emergency escape hatch. This is commonly referred to as a top-of-the-car exit. In passenger cars they are located at the back portion of the car. In many instances, it will be hidden above a false ceiling or light baffle, which you will have to remove to find it. At times, it will be opened from the inside of the car by a key. This is allowed under the seismic requirements in ASME A17.1. Remember, they are locked from the outside, but with a little *forcible* energy you will overcome it.

Review Questions

1. What is the COP?

2. Whose responsibility is it to maintain the 24-hour call center?

3. Describe the difference between independent and firefighters' emergency operation.

4. What key operates the fire operations panel?

5. How is the car top exit locked?

Field Exercise

With the members of your fire company, visit three buildings in your second due district and make note of the variations in the elevator car interior components.

Chapter 7

Hydraulic Elevators

History

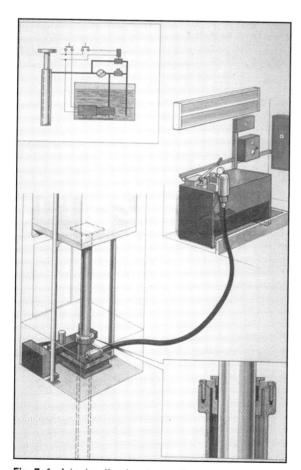

The earliest hydraulic elevators were water-powered (true hydraulic motion) freight elevators, which were later replaced by oil-powered hydraulics (fig. 7–1). The heyday of hydraulics came after World War II with the tremendous building boom across North America. During the next 50 years, 500,000 units were installed, both passenger and freight varieties. Until recently, most were in-ground design, which we discuss later in this chapter. Hydraulic elevator systems differ from traction systems in several distinct areas:

- Hydraulic systems do not require a penthouse or an area above the hoistway for equipment.

- Hydraulic systems are less complex and do not employ counterweights, car safety systems, speed governors, or hoisting ropes. An exception is the roped hydraulic, which is covered later in this chapter.

- They typically service buildings with heights of no greater than six or seven stories, and have a normal height limit of 60 feet.

- The cost difference benefit of hydraulic versus traction seems to equal out at the five stops or landings.

Fig. 7–1. A hydraulic elevator system

Taller buildings generally require traction systems due to their height and the need to move more passengers quickly. Hydraulic elevators travel at speeds of about 200 feet per minute (ft/min) or less, unlike their traction counterparts, which can travel up to 1,000 ft/min or greater in very tall high-rise buildings (fig. 7–2).

feet of the hoistway wall, but in some cases the distance is greater. There is no worse feeling for firefighters than to search endlessly for the machine room, only to find they were either standing beside it, or it is so far away that it is beyond comprehension. Preplanning is the key. Get out into your buildings and review all elevator components with building maintenance or elevator service personnel. In the case of the remote hydraulic installation, numerous safeguards would be required of the installer, to assure quick access by the mechanic.

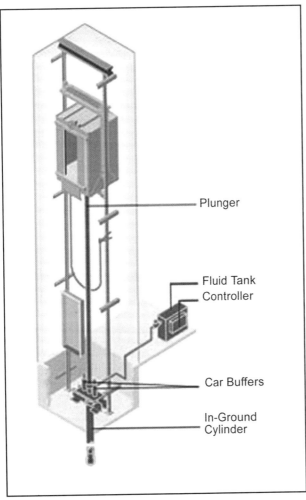

Fig. 7–3. Hydraulic fluid supply tank

Fig. 7–2. In-ground holed hydraulic elevator (courtesy of Otis Elevator Co.)

Hydraulic Machine Rooms

The machine room location will typically be located at the lowest landing, but in reality it can be anywhere, on the roof, in a closet on a middle floor, and so forth (fig. 7–1). The ASME A17.1 Code does not set a distance requirement, but most will be found within 10

The hydraulic machine room has a supply tank of combustible hydraulic fluid, usually 200–1000 gallons or more, that often becomes the source of fire alarm activations in buildings (fig. 7–3). The submersible pump located in the hydraulic fluid supply may malfunction, and the fluid can become overheated to the point of generating smoke. When this happens, the smoke detector in the machine room goes into alarm, bringing us to the scene. Unfortunately, many of the covers to the supply tanks will be left open or off, to allow dissipation of heat from the fluid, which is a violation of the ASME A17.1 Code. One of the major problems with hydraulic operations is the loss of function during very hot or cold days, with breakdowns commonly due to the loss of viscosity (resistance to flow) from the heat or sluggishness due to the cold (fig. 7–4).

readily accessible, while others present great danger to the firefighter because of their location in the pit. This valve should only be applied if mechanics already had secured the car in other ways.

Fig. 7–4. A hydraulic pit, with scavenger can, spring buffers (200 ft/min capacity), cylinder and plunger follower and shutoff located on supply line. *Do Not Enter!*

Fig 7–5. Shows a hydraulic elevator that fell when the supply line became dislodged at the cylinder. The car fell, then struck and injured a mechanic who was working under it without protection being installed, which is a violation of industry safety practices and procedures. The car is resting on the cylinder flange. There is no room!

During an emergency involving a hydraulic elevator, it will be necessary to shut down power and perform lockout/tagout. This would occur during a normal elevator problem in which we were removing a passenger who is not in any danger, just "locked in the box." *Keep in mind that the unit can still move if there is a leak in the hydraulic system.*

According to the *Elevator Industry Field Employees' Safety Handbook,* section 7(h), before getting under a hydraulic car, a mechanic must install jack stands, pipe stands or other approved scaffolding to protect him from a falling elevator (fig. 7–5).

However, if we are involved with a rescue of someone pinned or crushed by the elevator, then we must make every effort to further immobilize the car. This could be done at a number of points within the system by operating valves, under the direction of an elevator mechanic who will shut off the movement of the hydraulic fluid. During a crush-type incident, the car will be pushing in a downward direction, and slippage of fluid past the seals would undo any extrication effort. The work that you might have done over 30 minutes could be lost by the "creeping" of the car as a result of the liquid slipping past the seal. Some of these valves are

In figure 7–4 note the valve handle on the fluid line, between the cylinder and the spring buffer. An easier valve to operate is located as the fluid supply line leaves the supply tank in the machine room. Others may be located on the supply line as it makes its way from a remote machine room to the actual location of the elevator (fig. 7–6). The pictures show a valve with a handle on the line that would perform the "locking up" of the fluid from that point upward into the system, unless you are dealing with a serious leak beyond that point (fig. 7–7).

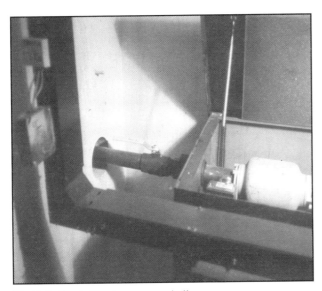

Fig. 7–6. Shutoff valve on supply line

Fig. 7–7. Hydraulic pump

A serious problem for the entire hydraulic industry is the failure of numerous single bottom cylinders without a safety bulkhead located in the ground that were installed prior to 1971. The terrible damage caused by the electrolytic or galvanic action on the metal has been the cause of many catastrophic failures of these cylinders. It was reported in *Elevator World* in 2001 that up to 3,000 in-ground cylinders required replacement per year because of serious leaks, both pinhole and major.[1] The fire service should note and recognize the seriousness of this failure rate and its potentially deadly consequences. In the past 10 years,

millions of dollars have been spent in litigation because of injuries and subsequent deaths resulting from single bottom hydraulic failures. The outlook is that passenger hydraulics may be limited to short-rise installations, while large capacity freight hydraulics will continue to be built. This type of design, cylinders without a safety bulkhead, was prohibited by ASME A17.1–1973 and later editions of that code. ASME A17.1–2000 and later editions require that hydraulic elevators without a safety bulkhead that are in the ground have their cylinder replaced with a cylinder having a safety bulkhead or that the elevator be provided with safeties. (See chapter 9, New Technology.)

It seems that the machine room-less (MRL) elevators that we see going into service every day will heavily impact that market, except for installations that are not in the ground. They will not replace the entire hydraulic market; freight elevators will continue because of their design and load-lifting capability. However, it will impact those who would have considered purchasing a direct-plunger hydraulic passenger elevator with an in-ground cylinder. They will either have to install a double-bottomed unit, with a PVC (polyvinyl chloride) liner to protect the metal cylinder, or go with telescoping pistons or some other above-ground design arrangement. These requirements have been in the ASME A17.1 Code for many years. If your authority having jurisdiction (AHJ) has not adopted the ASME A17.1–2000 or later or ASME A17.3–Existing Elevators Code, then the change is not mandated.

A family of hydraulic elevators is shown in figure 7–8. The family is varied, as you can see, with roped (2-inch lift for 1 inch of piston), piston (single or multiple units), telescoping (many in one) and in-ground varieties (see fig. 7–2).

As instructors (and both of us having served our communities for over 35 years), we caution all firefighters regarding the lowering of a hydraulic elevator. In chapter 11, Elevator Safety: Philosophy and Principles, we stress the four principles of safety. We state that we never lower or move an elevator without a mechanic performing that function. The elevator has stopped for a reason unknown to us, and it takes the trained mechanic to determine whether or not to move the elevator. ASME A17.1 (Safety Code for Elevators and Escalators) and A17.4 (Emergency Evacuation Guide) prohibit anyone being given instructions to lower an elevator other than a trained elevator mechanic. The

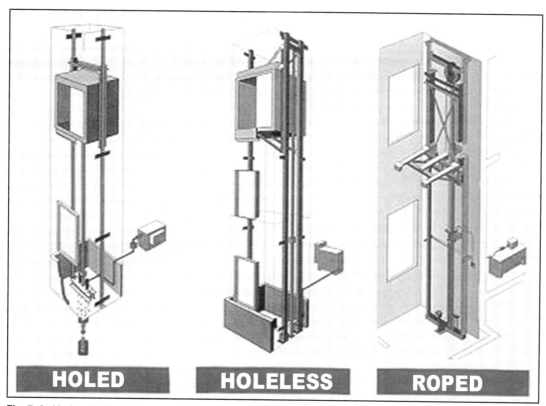

HOLED   **HOLELESS**   **ROPED**

Fig. 7–8. Hydraulic family (courtesy of Otis Elevator Co.)

courts will find you and your department had worked beyond their level of training when things go wrong. *Our advice: Wait for the mechanic, or if an emergency situation, power down, use lockout/tagout, gain safe entry to the elevator car, and remove the passengers.* For further information relative to the MRL elevators and their impact on the elevator industry, see chapter 9, New Technology.

Roped Hydraulic

The roped hydraulic has created an area of confusion for firefighters. How can it be a hydraulic if it has a rope and a sheave wheel in the hoistway? The design of the roped hydraulic provides a ratio of 1:2, where a piston with a stroke of 50 feet can provide a lift of 100 feet. The wire ropes are passed over a sheave located on top of a vertical jack located either beside or behind the car. As we know, it doubles the distance attainable for the designer. They also include a rope safety and governor in the design. In 1955, no major manufacturer considered that roped hydraulic had any future, and it was dropped from the ASME A17.1 calendar. By the

1980s, it was noted that they were being used in Europe successfully, and requirements once again appeared in the ASME 1989 Addenda to the ASME A17.1 Code. As the roped hydraulic is hung rather than supported, car safeties are required. The roped–hydraulic driving machine includes the cylinder, plunger, and sheaves and their guides (fig. 7–9).

Fig. 7–9. Roped-hydraulic cylinder/sheave

Telescopic Jack

The telescopic jack is used in some installations to increase travel of holeless hydraulic elevators (fig. 7–10). This type of jack is also used on roped hydraulics, and stages of the jack are synchronized so they move at the same speed. This function is similar to older car antennae, where the smallest dimension is at the peak of extension. They then "seat" down into the next larger dimensioned segment. They are all connected to each other to prevent any section from getting ahead of the other.

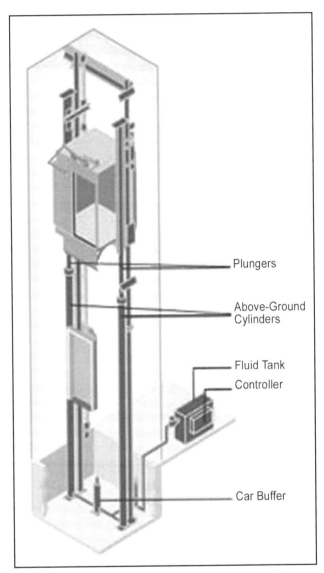

Fig. 7–11. Holeless hydraulic elevator (courtesy of Otis Elevator Co.)

The dual-piston hydraulic elevator has a jack on either side of the car, attached at the top of the car (fig. 7–11). The pistons push up to raise the car, then lower it as the oil is released into the supply tank (fig. 7–12).

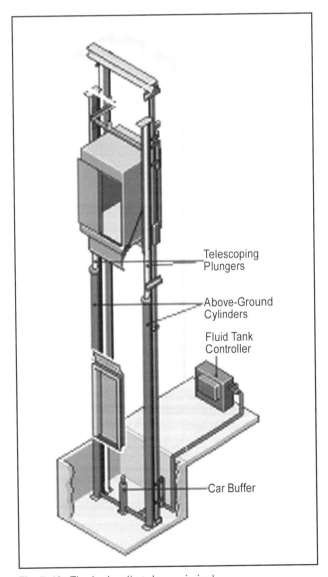

Fig. 7–10. The hydraulic telescopic jack

Fig. 7–12. Piston in action

Fig. 7–13. A hydraulic piston that shook! The piston installation was seen in Minneapolis, Minnesota.

Fig. 7–14. A plunger follower guide is installed.

As the car went up, the shaking and noise were very noticeable (fig. 7–13). In figure 7–14 a plunger follower guide is shown. They function as an auxiliary car plank to provide lateral stability. The guide is picked up from a stacked set depending on the height and design of the system. Sometimes they are found when a modernization is done and the existing hoistway dimensions do not provide a tight or secure enough ride for the passengers. If these had been installed in figure 7–13, the ride would have been quite different.

Summary

The world of hydraulic elevators is one that is filled with many styles of operation. The fact that there are 500,000 units installed across North America is reason enough to draw our attention. They include the following types:

- In-ground cylinder
- Above-ground (holeless)
- Dual-piston both above- and in-ground
- Telescopic piston
- Roped hydraulic
- Freight

Remember, that there have been reportedly up to 3,000 replacements of hydraulic cylinders yearly. These are mostly for minor pinhole leaks but require replacement of the cylinder. Do not forget that the supply line is also a very vulnerable part of the system, which, when damaged, can cause a sudden loss of hydraulic fluid and the pressure to keep the car under control. Never stand on the supply line, because the resulting damage could be reason for the sudden movement of the car into the pit.

Hydraulic elevators will continue to be marketed, but they will be seriously challenged by the MRL industry as replacements to their share of the market. The exception will be in the freight elevator area, where their lifting capacity is a plus for them.

Finally, we *never* lower an elevator, of whatever design or means. This is a job for trained elevator mechanics who must examine safety points before they lower the car.

Fig. 7–15. What type of elevator?

Review Questions

1. How high do hydraulic elevators usually travel?

2. Name the various types of hydraulic elevators.

3. Describe the basic differences between a traction system and a hydraulic system.

4. What effect does heat or cold have on a hydraulic elevators performance?

5. Identify the type of hydraulic elevator shown in figure 7–15.

Field Exercise

Locate 10 hydraulic elevator installations within your first-alarm district. List them as to their type, height, machine room location, and power disconnect locations in the machine room.

Endnotes

[1] Koshak, J. "Falling Hydraulic Elevators in Need of a Safety." *Elevator World* (December 2001), p. 101.

Chapter 8
Freight Elevators and Dumbwaiters

Freight Elevators

Freight elevators are an entire system of their own. The scene in figure 8–1 is a truly formidable obstacle to a rescue scenario, unless you have a good familiarization with a freight door system. To be brief, they are big, slow elevators that are used to move freight and those who are operating it. Some are monsters, capable of moving entire tractor trailer truck units into exhibition halls (see fig. 8–12), whereas others are small units tucked into a corner of a 150-year-old mill building forgotten by everyone. Earlier units were water-hydraulic powered, and others were part of the belt system that powered all of the machines in a mill. This was followed by the development of basement drum machines, then hydraulic and traction machine varieties serving the needs of industry. They have had a long role in the elevator world, being the original reason that elevators were designed: to move goods or materials from point A to point B.

Fig. 8–1. A power-operated biparting freight door system

Their history is a checkered one, in that a major number of the fatalities and injuries associated with elevators have occurred around the earlier versions of freight elevators. In the past, they were the movers of all products, finished and raw, in the early mill-constructed buildings, some of which are still operating after 100 years. The *shipper rope* elevator was operated by pulling up or down on a rope, usually manila, that ran through an elevator car. It was familiar to anyone who worked in older buildings, with only a *bar gate* protecting the hoistway opening. This left an open, unprotected hoistway, which claimed many victims in falls and crush injuries, and these types of elevators probably still exist today in some buildings. These installations presented many fire departments across the country with some of their most challenging extrication operations, commonly with a person, many times a teenager, caught between the front of the cab and the sill plate of a landing. Why young people? It usually was their first job, working in a factory, with no life experience or knowledge of hazardous conditions. These open elevator hoistways were also responsible for the spread of fire and smoke in many fatal fires via the unprotected openings onto other floors in the fire building. Although many still exist today, ideally they have had floor doors, interlocks, and car gates added to protect people from the hoistway, automatic operation, and other features that we find in the modern freight elevator.

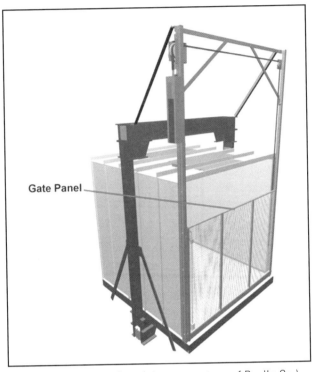

Fig. 8–3. Car gate and car (photos courtesy of Peelle Co.)

Figures 8–2 and 8–3 show a set of freight drawings of a modern system with floor doors, car gates, and so forth. This happens to be a hydraulic unit, but you will find them in traction applications as well. Figure 8–4 shows a freight elevator with a set of power-operated biparting doors with a safety astragal door edge that provides hand protection, and an open-mesh type gate serves as the car gate. This type of elevator is operated as a passenger car is, with the selection of the floor being entered onto the car-operating panel. The unit in figure 8–5 has a manual set of biparting doors, with a mesh type gate. Note that as you push the lower door down, it becomes part of the car/floor threshold. As you push down, your force actually will raise the upper door panel, which counterbalances the bottom door panel the same distance up. The lower panel is connected to the upper with a system of chains and pulleys.

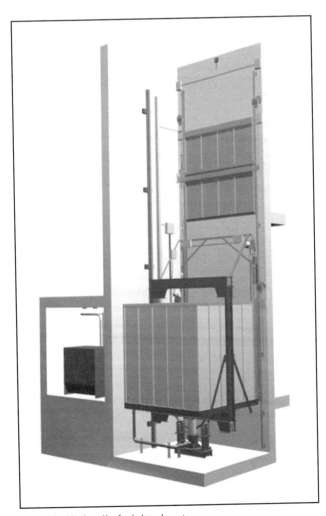

Fig. 8–2. Hydraulic freight elevator

Fig. 8–4. Typical larger door (courtesy of Peelle Co.)

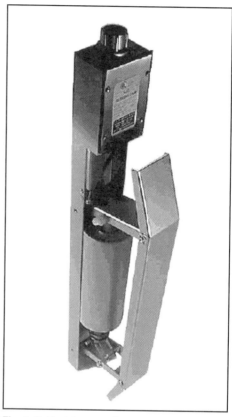

Fig. 8–6. Retiring arm (courtesy of GAL Manufacturing Corp.)

Fig. 8–5. Biparting manual freight doors

Figure 8–6 shows a retiring arm, and figures 8–7 and 8–8 show *cam* devices. In the operation of these units, as the car moves into the landing or unlocking zone, it pushes the interlock rollers to open the interlock or mechanical lock and electric contact, which is similar to an interlock and permitted on low-rise freight installations. The retiring cam arm is located on the outside of the elevator car, at the upper portion of the car wall. This allows the biparting doors to be opened manually from either outside, on the landing, or from inside the car by the occupant after lifting the manual car gate.

Fig. 8–7. Swing cam (courtesy of Peelle Co.)

Fig. 8–8. Double swing cam on a center-opening door

A power-operated gate can be opened by the car door gate operator, located on top of the elevator. In earlier installations, there were always stationary arms, which primed the interlock to the open position every time the car went by a floor. Eventually came the development of the retiring arm, which only comes out when the elevator approaches the selected floor, and the cam engages the retiring arm. This also eliminated the "stealing" of the elevator by others as the car went by their floor, and the ensuing heated verbal battles about who had the elevator first. Figure 8–9 illustrates the interlock with its release cam in position, awaiting operation by the retiring cam on the elevator. A mechanic is simulating the operation with his hand pushing on the cam. Figure 8–10 shows a freight cam and interlock, and figures 8–11 and 8–12 show two different freight doors.

Fig. 8–10. Freight cam and interlock

Fig. 8–9. Freight cam manually activated

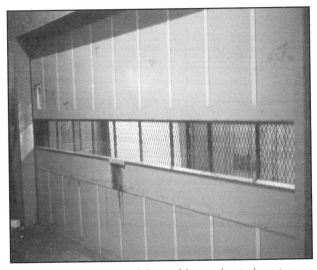

Fig 8–11. Power-operated door with panel gate (courtesy of Peelle Co.)

Fig. 8–12. Very large biparting door system

Fig. 8–13. An older unit

Emergency top exits in freight elevators usually run across the full width of the car with a minimum depth of 400 millimeter (mm), or 16 inches. Usually they are located in the front of the car top enclosure. There are many freight cars that do not have any ceilings; they are open to accommodate tall freight, to allow fully loaded tractor trailer units to access the convention floor from the street to facilitate setup of product or demonstration materials. This is commonly found in museums, convention centers, and similar type of structures. If they are located in your response district, drop in and ask management to give you a familiarization tour *(prefire planning!)*. By getting out into your district you will know what to do when you run into circa 1920 equipment.

Authorities having jurisdiction (AHJs) differ in how they treat older equipment. Some do nothing, whereas others have mandated minimum requirements such as the Safety Code for Existing Elevators and Escalators, ASME A17.3 (fig. 8–13).

Figures 8–14 and 8–15 show wooden floor doors and a panel gate freight, respectively.

Fig. 8–14. Wooden floor doors

Fig. 8–15. Open floor doors, showing panel gate freight

Dumbwaiters

The dumbwaiter machine is one that is not intended for the transport of humans. Unfortunately, the fire service usually meets the dumbwaiter when a person tries to ride one of these material handlers and ends up in a crush emergency. Dumbwaiters are simple machines meant to transport products from one floor to another. They are found in various height configurations and in many varieties of occupancies. In earlier times, large homes had dumbwaiters, in which food products were sent from the kitchens in the lower floor to the dining room upstairs. These were mainly of the hand-operated counterweight system, with no power-driven machinery. The kitchen help simply pulled on the rope to bring the car to their level, placed the goods into the open car, and then pulled or pushed the rope in an up or down direction. As their use grew, it was common to find them in restaurants, hotels, dining commons of schools, and many other places. They were of many designs, including traction, hydraulic, and drum machines. Figure 8–16 shows a two-car system with its floor doors closed. In figure 8–17 the floor doors are open when the car(s) arrive, and the floor doors are manually opened.

Fig. 8–16. Dumbwaiter floor doors closed

Fig. 8–17. Dumbwaiter car doors

Fig. 8–19. Stationary arm and retiring cam

Fig. 8–18. Car top

Figure 8–18 shows the top of the car. Note that, unlike the elevator, there is no emergency exit available on top of the car to assist the firefighter. Remember, these are material handlers, not meant for human occupancy.

In figure 8–19 a retiring cam is shown engaged to allow the opening of the floor door.

Fig. 8–20. Extrication attempt

Fig.8–21. The car the victim rode in

In figures 8–20 and 8–21, the dumbwaiter is shown that a university student attempted to ride in to gain access to the kitchen in a central dining room. Having no familiarity with the dumbwaiter system. the rescuers attempted to gain access to the student, whose arm was hanging out of the small vision panel in figure 8–20. Inside, he was pushed down by the upward movement of the box, but his arm was still hanging out of the vision panel. He was still in the car, with his arm jammed out from the vision panel, when the mechanics who had been called by the fire department arrived. The mechanics moved the car via the small electric motor used to run it, taking the pressure off of his arm, which allowed him to pull it back into the car.

An important point to remember from this incident is that it is vital that the fire department dispatchers notify the elevator company that it is a *crush injury*, and not a "stalled elevator" or a "man is stuck in the elevator." By clearly describing the incident as a crush injury, the proper help necessary to assist them will be dispatched immediately.

Summary

The history of the freight elevator is one that is a reflection of the progress of the industrial trail of the continent. Wherever there was shipping, manufacturing, or farming, there were goods that needed to be moved. After all, that is the original purpose of the elevator, to move goods, not people.

The early freight elevators comprised many varieties and included those powered by steam, water pressure, belt-driven, basement drum machines, traction, and hydraulic. They were the scene of many injuries and fatalities during their early development, and unfortunately that history continued as they are often still used in old buildings that are being reused for new industries.

The doors for early freight elevators were sometimes nonexistent, usually consisting of just a bar gate across the opening at the hoistway. Now they have the hoistway opening protected by a vertically sliding door, car gates, and an interlock. The doors are usually the vertically biparting or vertically rising type, and have a retiring or stationary cam and interlock arrangement. Remember that there are ancient machines still in existence, for example, in the back section of a mill building, that no one knows exist. The owners have their own people do the repairs and maintenance, and the fire service will not know about it until they roll into an accident involving the equipment and an employee.

In addition, freight elevators with manual doors have no ability to automatically respond to Phase I recall. If they are left standing at a floor with the doors open, they will not return to the designated or alternate level until someone manually closes the doors. This can create an accountability problem for firefighters trying to identify all the elevators upon arrival.

Review Questions

1. Describe the various types of drive power used in freight history.

2. What type of door system is usually found in a freight elevator?

3. What were freight elevators primarily developed to move?

4. How many floors does a dumbwaiter usually travel?

5. What is a *crush injury*?

Field Exercise

Locate the nearest freight elevator to your quarters, and create an index card system that can be retrieved by your dispatchers during an emergency involving this unit. This index should list all of the necessary information that a firefighter will need to retrieve, such as:

- Elevator service company emergency numbers

- AHJ inspectors' phone numbers and addresses

- Machine room location and disconnect location

- Owners' and managers' numbers

Chapter 9
New Technology

The elevator industry is being continuously exposed to new ideas and designs, as each elevator company pursues a better share of the marketplace. We could never quite keep up with the continual flow of ideas and equipment in the field today. In this chapter we will familiarize you, the firefighter, with some of the more important ones that you are likely to encounter.

Pictured in figure 9–1 is the Firehouse Inn (now The Kendall Inn) near the campus of Massachusetts Institute of Technology in Cambridge, MA. It became the site of the first machine room-less (MRL) elevator installed in Massachusetts during a pilot program authorized by the Massachusetts Board of Elevator Regulations. For the first 100 years of its life, it housed the quarters of the now shuttered Engine Co. #7 of the Cambridge Fire Department. The smaller building in the front of Figure 9–1 was the firehouse, and the rear section is the new addition, which includes the MRL elevator.

Fig 9–1. Site of pilot installation of MRLs in Massachusetts

Machine Room-Less (MRL) Elevators

As the latest member of the elevator family, the MRL is actually not a new creature; it has been on the market in other areas of the world for 10 years or more, pioneered by KONE Corporation, and with newer entries in the market by Otis Elevator, Schindler, ThyssenKrupp, and Fujitec. As you read in chapter 7, the future of the direct-plunger hydraulic passenger elevator with an in-ground cylinder is facing serious potential environmental issues. The MRL market is attractive to business owners looking for replacement for several reasons, including the following:

- The excessive noise and odor of hydraulic units

- Hydraulic units' poor performance in hot and cold environments

- Environmental impact of hydraulic fluid spills

- Electrolytic erosion of single-casing, in-ground cylinders, causing catastrophic collapses (This is not the case with installations that comply with A17.1.)

- The cost of pulling those cylinders, inspecting, and replacing them.

Another driving force behind this push for the MRL elevators is the opportunity to have more rentable space available to the prospective building owner. Where rentable space is a premium, this means more money for the owner. Other big fans of this design are architects, because they view elevator penthouses as ugly protrusions jutting up from their graceful buildings. All of the previous reasons, as well as the availability of newer, lightweight motors and equipment and the new codes addressing the safety of these systems, have contributed to this surge in the MRL field.

The concept of the MRL is that all of the mechanical drive equipment is located within the hoistway walls. It is usually mounted in the upper portion of the hoistway on a side wall behind the guide rail (KONE), adjacent to the guide rail (Fujitec), directly over the car (Otis,

ThyssenKrupp, Schindler) on a beam structure, or in the pit directly under the counterweight (ThyssenKrupp). The controller and mainline disconnect switches usually are located in a control space, or small room (e.g., a closet) adjacent to the hoistway at the top or the lowest floor of the building. Each company accomplishes any necessary hoistway maintenance by placing the elevator on inspection operation with most hoistway work being done by the mechanic while on top of the elevator. The different companies have distinctly different systems to avoid patent infringement. In addition to the location of the machines and other equipment, the suspension systems are dramatically changing.

Otis Elastomeric Coated Steel Belt (ECSB)

Otis uses a polyurethane (PU) covering over high-tensile-strength, zinc-plated, steel-corded drive belts for traction, rather than the steel hoist ropes (see fig. 9–3) that we usually expect with the traction elevator. This type of belt is listed as an elastomeric coated steel belt (ECSB) in the ASME A17.6 Standard for Elevator Suspension and Governor Systems.

ThyssenKrupp Aramid Ropes

ThyssenKrupp and Schindler (Schindler in Europe only at present) use a new hoist "rope" made of Kevlar (aramid) material in their MRL roped elevators. (Kevlar is a registered trademark of DuPont Chemical, and aramid is the generic chemical name of Kevlar.) KONE uses standard steel wire hoist ropes with a permanent magnet (PM) motor gearless machine. Fujitec has two pressure-sheave wheels that create an indirect drive system where pressure on a belt made of aramid fiber is applied to the traction sheave for motion.

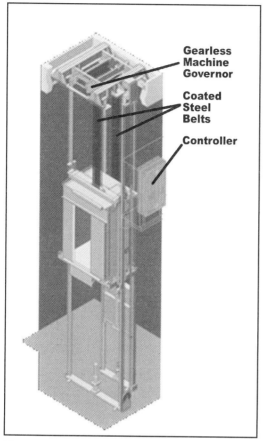

Fig. 9–2. Gen2 machine room-less (MRL) elevator by Otis Elevator Company

The Gen2 made by Otis Elevator is an MRL elevator system that places the machine, brake, and overspeed governor in the hoistway (fig.9–2). The controller is mounted in a control room or space (closet) next to the hoistway, usually at the top landing. It provides a means of remotely operating the elevator and performing tests on the brake and governor, which are located in the hoistway and are much less accessible to service personnel than they were when they were located in a machine room. The controller also provides a means to lift the brake (*for elevator mechanic use only*) to allow gravity to move the car. This means is available to rescue trapped passengers during a power outage or equipment failure, or to raise the car to the top of the building to access the machine, encoder, governor, and brake in the event of a power failure.

The control room (or space) is located at or near the top landing adjacent to the hoistway. The suspension means consists of five 30-mm polyurethane (PU)–coated high-tensile-steel belts for the limited duty product release, or three to five 60-mm belts for the extended duty release (fig. 9–3). Each belt contains 12 to 24 steel cords (30- or 60-mm belts). These cords are laid out flat and encased in polyurethane. The breaking strength of each belt is guaranteed to be a minimum of 32 kN and 64 kN for the 30- and 60-mm belts, respectively.

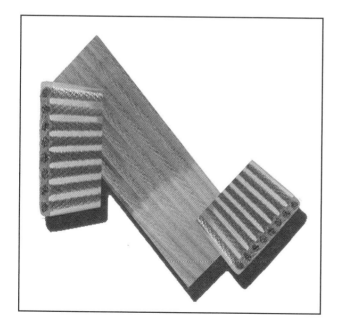

Fig. 9–3. Coated steel belt (photos courtesy of Otis Elevator Co.)

Fig. 9–4. Otis Gen2 machine and support frame

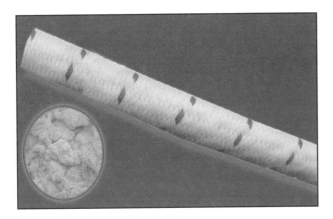

Fig. 9–5. Aramid rope

Fig. 9–7. Steel wire rope

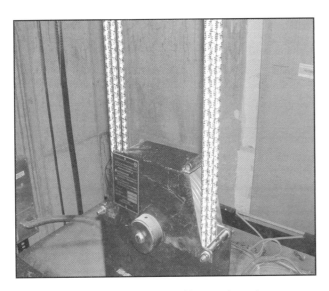

Fig. 9–6. ISIS-1 sheave and aramid ropes in a pit

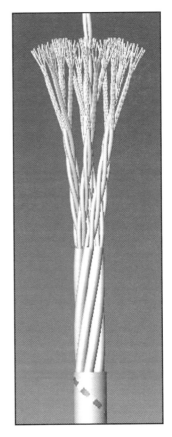

Fig. 9–8. ISIS-1 K-Core rope

The ISIS family of products is the entry into the MRL market from ThyssenKrupp AG (fig. 9–6). All the mechanical components are located within the hoistway itself, with the controller and power disconnect in a machine space or closet adjacent or near to the hoistway.

The major differences are that instead of a steel wire rope for traction (fig. 9–7), hoist ropes are used, made of aramid fiber, called the K-Core (fig. 9–8). ThyssenKrupp is also using nonmetallic sheaves made up of composite materials and, in low-rise applications, locates its geared machine in the pit of the hoistway, thus requiring the mechanic to get under the car to accomplish specific work. This is not new, as mechanics have had to work in the pit before, but never on a machine. They have means of blocking the car so that it cannot come down on the mechanic, but if it is not used, this creates a very dangerous situation. (See fig. 9–9, ISIS-1 machine located in the pit.) Starting October 1, 2005, ISIS-1 is being phased out of the market and is being replaced with ISIS-2 systems in most cases. There are more than 500 ISIS-1 systems installed in the United States and Canada, and they will stay there and be serviced when under contract.

Under no circumstances should a firefighter go under any elevator car until the power is removed from the elevator, and it is completely immobilized with the proper blocking. A particular danger with hydraulic units is that the hydraulic lines in the pit may break when accidentally stood on during the extrication, bringing the car down instantly. Do not add us to the list of dead and injured for the incident because we rushed into the situation.

KONE Mono-Space MRL System

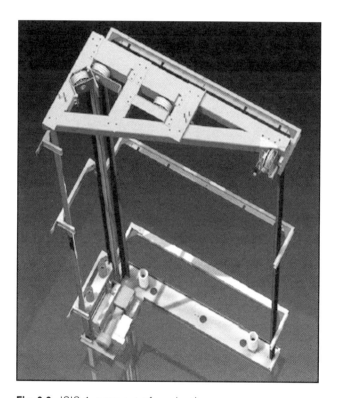

Fig. 9.9. ISIS-1, now out of production

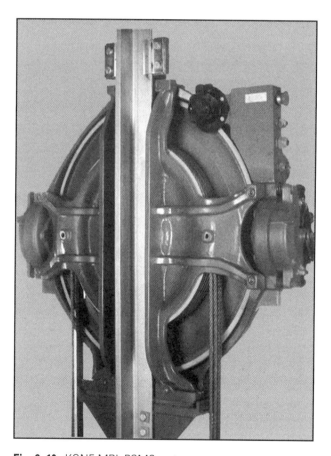

Fig. 9–10. KONE MRL PSMS motor

The ISIS-2 uses an overhead beam structure to mount the gearless machine, and it uses K-Core rope. All MRL systems provide blocking means, whether the machine is in the pit or the overhead. This system employs a blocking means that ties the car to the structure so that maintenance and repairs can be carried out safely on the top of the car.

Fig. **9–11.** KONE control closet next to hoistway door

Permanent Magnetic Motors

The KONE motor shown in figure 9–10 is a Mono-Space unit, with a limit of 10 landings. A larger version, called the Eco-Space, is being installed in buildings up to 36 stories. (Fig. 9–11 shows a KONE control closet next to a hoistway door.) This slim-line permanent magnetic synchronous motor (PMSM) is located under the guide rail at the top of the hoistway (fig. 9–12). All the mechanical parts of this unit are in the hoistway. The controller (below) and the main power disconnect (top-left) are located in a machine space (closet) adjacent to the hoistway at the top landing. The actual controller panel is the larger area, and for safety and maintenance reasons, it should be locked with an elevator company key to prevent tampering (fig. 9–13).

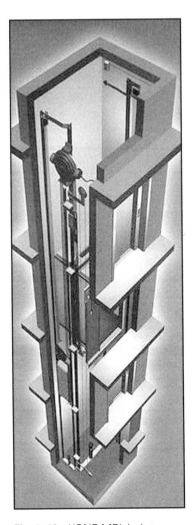

Fig. **9–12.** KONE MRL hoistway
(courtesy of KONE Elevator)

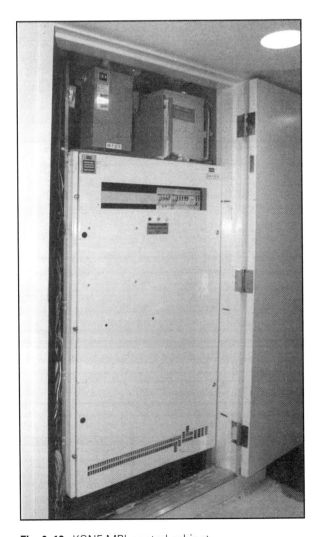

Fig. **9–13.** KONE MRL control cabinet

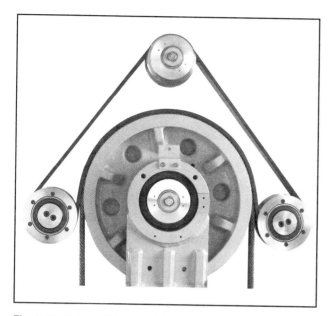

Fig. 9–14. Fujitec Talon machine

Fujitec Belts

Fujitec Talon is the name of Fujitec's entry into the MRL market (fig.9–14). In the Talon system, traction is provided by an aramid fiber belt imparting pressure on the hoist rope surface of the suspension sheave. This pressing force transmits the mechanical power from the motor shaft directly to the ropes. The differential weight of the car versus the counterweight is driven by traction at the belt rope surface area. The Talon drive mechanism consists of a drive pulley located at the top centerline of the machine that is directly connected to the slim permanent magnetic (PM) motor. Two additional drive pulleys are located on either side of the suspension sheave (fig. 9–15). The belt and the suspension sheave are located at the center of the drive mechanism. When the motor starts, the drive pulley transmits power to the belt, and the belt imparts motion to the ropes.

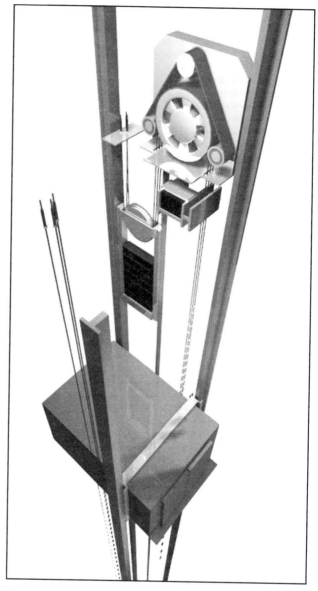

Fig. 9–15. Fujitec hoistway view of Talon system

The Schindler 400A may come in one of three designs: the MRL, MRS (machine room–side) and MRA (machine room–above). Note the location of mainline disconnect (right side of closet) and elevator to the right side of the control and drive cabinet in the control closet (fig. 9–16). We, as firefighters, have no business touching anything in this closet except for the mainline power disconnect.

The Schindler 400A is run by a compact gearless traction PM motor unit that is mounted in the top of the hoistway and fastened to the top of the guide rails by a machine-and-ropes termination support (fig. 9–17). The overspeed governor is also mounted in the same manner. All access for maintenance and inspection is done from the car top, and the car is immobilized by means of a lock-and-block device that allows the car to be suspended in the hoistway while the work is being done. This application is used in both the MRL and the MRS installations. The lock-and-block feature will be found in all MRL elevators on the market.

Some points about the 400A follow:

- May serve up to 20 floors, with 21 openings

- Travel of 108 feet (MRL) to 200 feet (MMS/A)

- Speeds of 200 to 350 ft/min

- One to four cars in system

- Weight capacity 2,100 to 3,500 pounds

An important question about MRLs for the fire service to ask is. "Where is the machine room or space located?" Some authorities having jurisdiction (AHJs) have mandated that the location be posted at the first floor elevator entrance (fig. 9–18). This enables the fire service to know where it is located. Note the marking on the top of the door frame for this MRL. It should also be emphasized that not all local authorities have gone along with the machine space concept. Some, in fact, have mandated a control room, where the mechanic can work behind a closed door while performing maintenance procedures.

Fig. 9–16. Schindler MRL control closet

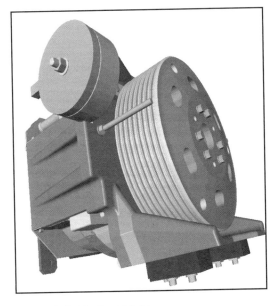

Fig. 9–17. Schindler PM drive

Fig. 9–18. MRL marking on door frame

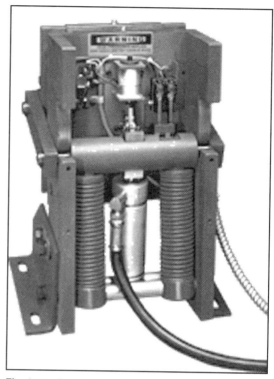

Fig. 9–19. Rope gripper 1

Safety and Rescue Equipment

The Safety Code for Elevators and Escalators has mandated protection against some specific failures:

- In electric elevators (roped or traction elevators), uncontrolled movement and ascending car overspeed must be prevented.

- In hydraulic elevators, protection against fluid and pressure loss causing uncontrolled movement must be provided in the form of replacing all single bottom cylinders or fitting them with a plunger gripper or safety.

Rope Gripper

The Rope Gripper™ from Hollister Whitney is used as an emergency brake, to prevent unintended motion of an elevator and to stop a car overspeeding in the up direction (figs. 9–19 and 9–20). The purpose of car safeties until now has always been to stop a car in a downward direction, set for either an overspeed in feet per minute (ft/min) or an excessive inertia setting. The Rope Grippers and similar products are meant to stop cars leaving the floor with their doors open with any unintended motion. They place a "grip" on the hoist ropes via a mechanical system, hydraulically set, that will stop the upward- or downward-moving car (depending

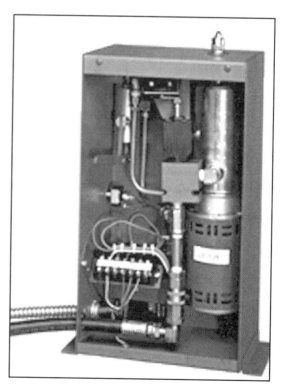

Fig. 9–20. Rope gripper 2

on the design) from crashing into the overhead or pit, and causing injuries and possibly deaths. The Rope Gripper is located on the hoist ropes and can be located in several different places, depending on the design of the system.

Ascending car overspeed and uncontrolled movement of electric elevators can occur in some failure conditions. To prevent this from occurring, the Rope Gripper is applied to prevent this movement. It can be mounted in the machine room or in the hoistway. It remains open around the steel wire hoist ropes until the controller indicates an unexplained movement of the elevator, and then the ropes are powerfully gripped, bringing the elevator to a safe stop.

GAL Rescuvator

The GAL Rescuvator and similar products provided by many manufacturers offer a means to safely lower a stalled hydraulic elevator car and release its passengers safely during a power loss. After checking the car's door and controller features, it lowers the car to the next lower landing and opens the doors to discharge passengers who might otherwise be trapped (figs. 9–21 and 9–22). The Rescuvator provides to the fire service a safe, reliable means to eliminate the continual response to that elevator call that we go to so frequently that we become part of the building's landscape. We are not being called to incidents that do not materialize since there is no entrapment.

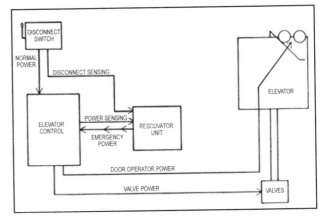

Fig. 9–22. Rescuvator system design

LifeJacket

The problem that the in-ground hydraulic elevator industry has encountered is that the older cylinders have catastrophic fluid and pressure failures, in the worst cases causing injury or death to the riding public. This is occurring primarily with those cylinders installed prior to 1971, as the code and engineering at that time only required a single-bottom cylinder, with limited corrosion protection. Being in the ground, the steel may be subjected to damage from electrolytic action, corrosion, or other causes, resulting in the sudden failure of the cylinder. Another cause of hydraulic catastrophic fluid and pressure failures has been the puncture or severance of the oil supply line to the cylinder. This has brought about the uncontrolled descent of the car into the pit, resulting in casualties.

After 1971, cylinders with a safety bulkhead were required, and in 1989 further changes to the code required polyvinyl chloride (PVC) or other protection covering the wall and bottom of the installations buried in the ground. To address the existing elevators in the ground, the A17.1 Code has adopted requirements that plunger grippers must be installed or the cylinder must be replaced. The LifeJacket® (made by the Adams Elevator Equipment Company; fig. 9–23) mounts around but is not in contact with the plunger in the elevator pit. It will automatically grasp the plunger when any uncontrolled movement including overspeed is detected. This will prevent the elevator from falling and causing injuries. Although a plunger gripper does accomplish a safety role, it may still be required that the cylinders be replaced when a leak develops in the cylinder.

Fig. 9–21. Rescuvator control box

Fig. 9–23. Adams LifeJacket

Fig 9–24. Smoke Guard Systems installation

Not all jurisdictions have mandated pulling the cylinders for inspection and replacement because the local legislation may not have adopted the latest codes. The use of the LifeJacket and similar devices is only putting off the inevitable—the replacement of the affected cylinders with properly protected ones, though in most cases the cost of a LifeJacket is significantly less than replacing a cylinder.

Smoke Guard Systems

Every day advances and changes are brought forth by industry, and figure 9–24 shows one of those. This lobby entrance at first appears no different from many others. We can easily note the smoke detector, automatic sprinkler protection, hall call buttons, and signage.

What makes this entrance different are the Smoke Guard Systems (Omega Point Laboratories) just above the elevator entrances (fig. 9–25). They are activated to roll a polyimide plastic sheet curtain down and seal off the car entrance when activated by the lobby smoke detector unit. This brand and others like it are being installed across the country with AHJ approval. Like many things we deal with, once you are familiar with the units, they will not present a problem, given that they are only held onto the door frame by magnets.

Fig. 9–25. Smoke Guard ready, and then deployed

The elevator industry is the scene of constant change. Manufacturers are constantly presenting new concepts and designs that they want to place into their repertoire and offer to the business world. This must be done through the ASME process, since most of the designs do not meet the existing version of the code. As firefighters, we will be constantly exposed to these new and sometimes unproven concepts. In the warehouses of industry lie the remains of the latest barn-burner product of last year. In some cases, products that never should have seen the light of day made it into the field and failed miserably, with resulting repairs required on an ongoing basis during their lifetime.

As with all we do, be aware and be safe!

Destination dispatch systems

A new technology that is becoming more popular is destination dispatch systems. The first was the Miconic 10 system (destination oriented elevator) introduced by Schindler, followed by ThyssenKrupp's DSC (destination selection control) (figs. 9–26 and 9–27). These represent a new concept in elevator traffic management, the elevator group central control system. Elevator engineers spend a great deal of time and energy figuring out the needs of a building and the occupants and meshing them with the timing and direction that the elevators will perform. As the passengers enter their destination on an input terminal in the lobby, the system will then assign them individually to a car that provides the most efficient journey time to arrive at the desired landing. In the car, a control panel will indicate the floor(s) being served and will flash when the floor is reached. There are no push buttons inside the car to allow you to change your requested floor. There are keypads in all lobbies to allow you to be picked up and delivered to your new destination.

Fig. 9–26. Schindler Miconic 10 hall touch pad

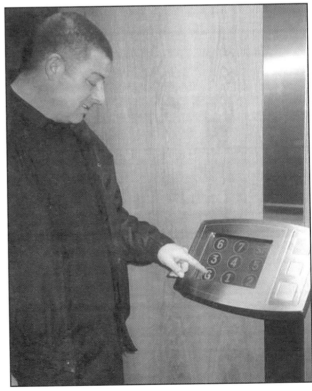

Figure 9–27. Destination selection control (DSC) in use

Phase II fire service needs are met by a panel that drops down or swings open to reveal a keypad or buttons for firefighters to use for floor selection. Phase II will operate the same as any other Phase II elevator. This style of group control is now in mid- and high-rise buildings, and is being installed across North America as this book goes to print.

Summary

As you have seen in this chapter, the changes in the world of elevators are coming at us fast and furious. Ten years ago, none of us in this country had even heard of the term MRL, never mind ever seen one. Now we have the following entries in the marketplace:

- KONE Mono-Space—The machine is behind the guide rail at top of the hoistway.

- ThyssenKrupp ISIS-1 and ISIS-2—The machine is in the pit and at the top of the hoistway.

- Otis Gen2—The machine is at top of the hoistway.

- Fujitec Talon—The machine is at the side of the hoistway at top.

- Schindler 400A—The machine is located at top of the hoistway.

These are followed by the new safety devices:

- Rope Gripper and other makes of uncontrolled upward-motion control devices are meant to stop unintentional motion with the doors open or overspeed in the up direction.

- The Rescuvator and versions made by other companies will, on the other hand, bring a car *down* to a safe stop during a control problem with a hydraulic elevator.

- The LifeJacket is another means to control a hydraulic elevator, but this is to prevent an *uncontrolled* descent down into the pit during a cylinder or supply line failure.

To think that a person would be able to select a floor, then be told which car to take to get there, was something we might have expected to see in a futuristic movie. The destination-oriented Miconic 10 and DSC dispatch system are doing just that in installations around the world. Your city is probably having them installed today.

The installation of smoke-separation doors is not new in many jurisdictions, but who ever thought that we would be allowing the sliding of plastic sheeting across the elevator opening to separate the hoistway and the building floor!

These changes and the ones yet to come are brought about by engineering advances and the demands of the economy to provide more for less. What we as firefighters must do is to be ever aware of the changes and advances that directly affect our daily working lives.

As firefighters, we are always concerned when we see plastics, which the fire service sometimes refers to as "*frozen gasoline,*" involved in anything in a building. In her book, *In the Mouth of the Dragon* (Avery Publishing Group 1990), Deborah Wallace points out the dangers that plastic presents to society. The Massachusetts Board of Elevator Regulations (MBER) has mandated that any new technologies being introduced into the hoistway be evaluated by an independent fire protection engineering firm, with the results of that testing forwarded to them. Will the ECSB burn? Will Aramid rope melt? What is the fire potential of *any* of the new technologies being introduced into the hoistway?

What was examined in the testing was the potential for flame development, fire and smoke spread out of the hoistway, and the toxicity of the smoke developed. The results of the tests were summarized and for further information contact the Massachusetts Department of Public Safety (DPS), Board of Elevator Regulations (MBER).

Review Questions

1. What problems are impacting the future of the hydraulic elevator industry?

2. ECSB means what, and what company does the product identify?

3. Describe the operational differences between a Schindler regular operating system and one with a Miconic 10 system.

4. What company has stopped production of its low-rise MRL with its machine in the pit?

Field Exercise

Locate five elevator systems in your district that are of the MRL family. Preplan their component's effects on a fire involving that system.

Chapter 10
Residential and Special Elevators

This chapter deals with a group of elevators that are considered special because of their appearance or use. As you will see, some have a terrible history that we have been involved with when responding to emergencies, while others are just strange to us because of where and how we find them. The elevators that we will cover are the following:

- Private/residential elevators

- Limited use/limited access (LULA) elevators

- Inclined elevators

- Double-deck elevators

- Observation elevators

We will try to stay focused on each particular unit, because they all do not have the same death and injury rates, but each one does possess inherent hazards to life and limb.

Private/Residential Elevators

Fig. 10–1. A modern private/residential elevator (courtesy of Otis Elevator)

The private/residential elevator is by definition as follows: Limited to a maximum load of 750 pounds, at a rated speed of 40 feet per minute (ft/min) and a rise of 50 feet. It is only installed in a single private residence (private home or a single residential unit in a multifamily building) or as access to a private residence, and is limited to no more than two adjacent landings. The private/residential elevator has been around the elevator business since the earliest elevator installations (fig. 10–1). The development of the "flying chair" in France from 1670 to 1750 is considered to be the beginning of the residential passenger elevator. A version of it was located in the palace at Versailles. It was a counterweight system used to transport the king's mistress to and from romantic liaisons.

As elevators were developed, it only made sense that it would be people of means who would be able to afford to have a private personal elevator in their homes. Sometimes these were just for convenience, while others were necessary to provide access to the upper floor(s) for those unable to maneuver the stairs. Many of us who have worked in cities have found these units in large older homes, which in their second life have sometimes become a local bed and breakfast inn. This is an illegal use for a private residence elevator, and it should be reported to the local authority having jurisdiction (AHJ).

Even though the A17.1–1955 edition code applied controls over this issue, a particular problem has troubled this part of the elevator market ever since earlier units were first installed. A space exists, at times in excess of 5 or 6 inches, when the hoistway swing door is closed, and the car gate is also in the closed position (fig. 10–2). This space or gap became an active danger with the onset of automatic push-button operation in the 1930s, as the elevator now could be called from a landing remote from the one that children were playing near. Children playing near this type of elevator have been severely injured or killed when the elevator moved. They are able to open the hoistway swing door, because the car is at their landing. Unfortunately, the car gate is closed with its safety circuit made up, except for the floor door interlock contacts. Once they close the hoistway door, the circuit is complete, and the car can respond to a floor call initiated elsewhere in the building. The child is standing on both the car and floor threshold at the

same time, and will be killed or terribly injured when the car moves up or down the hoistway. The tragedy of these accidents is that they are not isolated occurrences, but they are happening with predictable regularity. The modern elevator is designed not to have this gap space, as it is sometimes called (see fig. 10–1). It is eliminated in the design phase of the system, but a new hazard has been noticed in some jurisdictions. If an elevator company does not do the floor door installation, but an outside contractor is brought in to save money, the problem can creep back into existence due to a lack of appreciation for the danger on the part of the contractor to maintain the 5-inch *maximum* space between the floor door and the car door/gate and a 3-inch maximum space between the floor door and the floor edge of the landing sill. (See ASME A17.3–2003 Safety Code for Existing Installations).

Fig. 10–2. Private residence swing door elevator (courtesy of Otis Elevator)

Fig. 10–3. Small baffle at door bottom. This does *not* fix the problem. (courtesy of Otis Escalator)

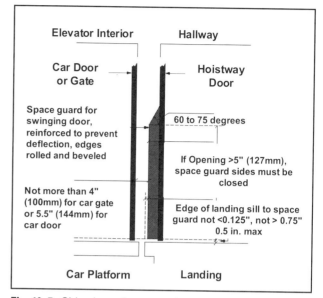

Fig. 10–5. Side view of entrance (courtesy of Otis Elevator)

A major campaign is underway to eliminate this gap space from the existing installations through an effort being spearheaded by Otis Elevator Company (fig. 10–3). The ASME Code at section 3.4.3 addresses the changes needed by requiring space guards (also called baffles) to be installed on the back of the swing door serving as the floor door. This will eliminate the gap hazard by filling the space on the back of the floor door, making it impossible for anyone to stand there when the door is closed. Otis offered to retrofit its customers' elevators with space guards at no cost (figs. 10–4 and 10–5).

It is clear that there are a number of impediments to this very commendable effort:

- Very few AHJs around the country have adopted *any* editions of A17.3, because the pressure from the business community is usually *not* to adopt, due to what they see as an unneeded burden on them to pay for meeting the requirements of that code.

- Complacency and the "my elevator is inspected every year and is safe" attitude.

- In the past, different versions of the baffle have failed, particularly because the building owners, maintenance crews, or mechanics have removed them *after* the installation for one reason or another.

- Finally, at the National Association of Elevator Contractors (NAEC) meeting held in April of 2005 in San Antonio, Texas, it was reported in *Elevator World* that fully 90% of new residential elevators are not inspected when they are installed. [1]

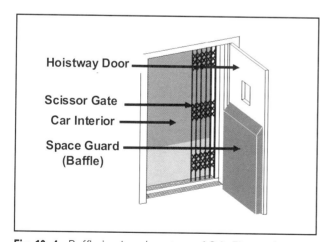

Fig. 10–4. Baffle in place (courtesy of Otis Elevator)

Shown in figures 10–6 and 10–7 is a manual side-slide door elevator located in an 80-year-old inn in northern New Hampshire. The shadow seen in the car door vision panel in figure 10–6 is the baffle installed long ago to address the problem with this elevator. With the sliding gate closed and the baffle in place, a child cannot fit into the potential space between the gate and the floor door.

The previous facts show us that this problem will not go away very quickly. The members of the fire service community can provide a useful set of eyes to help eliminate this life safety hazard. Check any swing door private residential elevator that you may come in contact with during your daily routine, and notify the local AHJ that a problem may exist in your district. Remember, *you* will be among the very first emergency responders to this terrible type of incident.

- A girl, age 4, was killed when caught between floors and an elevator in a residential building. Her mother had gotten off before her and other children pressed the call button. (5/1/97, Chicago, IL)

- A boy, age 8, died when he was crushed by a hotel elevator. He had become wedged between the elevator doors and a folding metal gate. (8/23/01, Bethel, ME)

- Two sisters, ages 6 and 7, were killed in a moving residential elevator. The elevator's safety feature was disabled, allowing it to ascend while the girls' heads stuck out past the gate. (7/31/02, Monmouth County, NJ)

(*Note: The disabling of any safety feature, on any elevator, can bring about the same tragic results*).

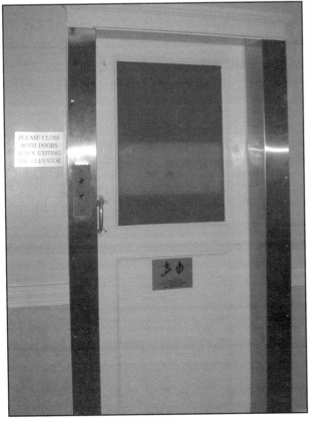

Fig. 10–6. Manual side-slide door

Fig. 10–7. Manual slide gate

Limited Use/Limited Access (LULA) Elevators

Limited use/limited access (LULA) elevators are specifically intended for installation in public buildings where no vertical transportation currently exists. Installation of a LULA can make a basement or second floor much more rentable. LULAs can improve access to houses of worship, restaurants and stores, and virtually any public facility

The majority of LULAs are *holeless hydraulic elevators*. Essentially, LULAs are small commercial elevators; they are typically provided with firefighters' service Phase I & II operation and illuminated call buttons. They are limited to 18 square feet of inside platform area and a rated load (capacity) of 1,400 pounds. LULAs are more cost-effective than commercial elevators for existing structures. LULA elevator equipment is less expensive than commercial elevator equipment, and the associated building requirements are reduced as well. Because LULAs are intended for installation in existing buildings, there are a lot of allowances made. The minimum clear overhead, for example, is specified as 12 feet; however, installation of a refuge space device allows the clear overhead to be reduced. ASME A17.1, section 5.2, has many of these allowances that permit the installation of LULAs in existing buildings.

Inclined Elevators

By definition, an inclined elevator is an elevator that travels at an angle of inclination of 70° or less from horizontal. These are found in hilly areas to provide access to the public who either do not want, or are not able to walk up or down a steep grade of considerable distance.

Double-Deck Elevators

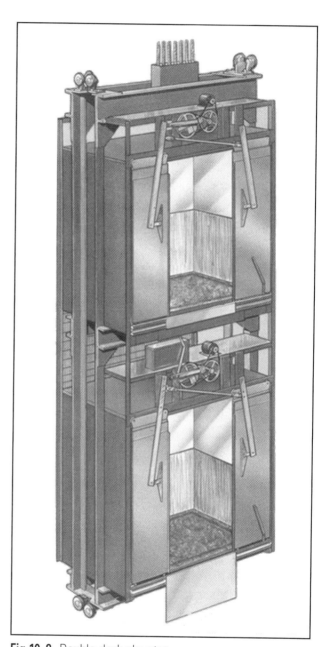

Fig. 10–8. Double-deck elevator

Double-deck elevators have been around since their installation for the first time in the 66-story New York Cities Service building in 1931 (fig. 10–8). The multideck elevator is an elevator having two separate platforms and compartments, one above the other and supported within the same car frame. At the present time, many of the first-generation installations are undergoing replacement or modernization programs. Both upper and lower decks may be loaded at the same time. Cars can be arranged to serve either odd or even floors, or all floors, whichever operation is desired. They are usually only found in mega-high-rise buildings due to the number of passengers requiring movement. There is an emergency top-of-car exit from the lower car to the upper car. The upper car also has a top-of-car emergency exit to allow elevator personnel to remove the occupants to the nearest floor door opening. The exit between cars must have protection for the evacuees to pass through to the other car. The Phase II operation by firefighters is conducted from the upper car, after the lower car has been emptied.

Observation Elevators

Fig. 10–9. Observation elevator

The first glass elevator was installed in the El Cortez Hotel in San Diego, CA, in 1927. The observation elevator as shown in figure 10–9 is found in most mall and atrium settings in the country today. They are located

in parking garages, shopping malls, hotels, transportation centers, and any other location where the architects can envision the riding public looking out over the building or city as it rises. Observation elevators are also put in not just for aesthetics, but sometimes for security reasons. People committing crimes do not like to be seen by a number of witnesses who can report the crime or take action such as calling 911 for assistance. Usually there are no top-of-car emergency exits from observation cars, due to the fall potential for all parties concerned. If the observation elevator is in a glass-enclosed hoistway, it will have a top emergency exit.

Summary

The elevator world is constantly developing systems to meet the needs of many aspects of the riding public. In this chapter we discussed five of those elevators:

- Private/residential elevators

- Limited use/limited access (LULA) elevators

- Inclined elevators

- Double-deck elevators

- Observation elevators

It should be remembered that each of these has a specific purpose and deals with a specific market. The private/residential elevator is exactly what it is called, an elevator to be installed in a private home. There are specific limits on its load capacity and number of landings, as well as a rise limit of 50 feet. The major concern with these elevators is the existence of a space larger than 3 inches in the area between the car door and the back of the floor door. This has turned out to be a killer of children in many instances. The baffle program sponsored by Otis Elevator is a campaign to eliminate those dangers. The LULA is a limited use/limited access elevator, which is exactly that. It is usually located in public buildings, with only people who have a specific need to use the elevator having access to it.

The inclined elevator is installed in locations of an incline of 70° or less, to provide a means of transporting people up and down a steep slope. A more familiar elevator in the larger cities is the double-deck elevator. These are usually only installed in mega-high-rise buildings, where there is a guarantee of heavy traffic flow to the upper floors of the building. Passengers select the car that will be stopping at their floor, odd or even floor number. Finally, the observation elevator has been on the scene since 1927. Today, they are found in many applications, such as malls, hotels, and other large and small buildings.

Review Questions

1. What is a baffle used for?

2. Can a LULA be placed in a new building?

3. Where was the first double-deck elevator installed?

4. Where is the top of car emergency exit on an observation elevator?

Field Exercise

In your second due district, locate and document two elevators of the following family: limited use/limited access (LULA).

Endnotes

[1] Gray, L. "A Brief History of Residential Elevators: Part 3." *Elevator World* (March 2005), p. 142.

Chapter 11

Elevator Safety: Philosophy and Principles

Introduction

Elevators are used by millions of passengers daily. Though equipped with many safety features and devices, elevators can suddenly stall due to mechanical or electrical equipment failure, building power outage, or human error. Injuries and deaths result from falls at the entrance to an elevator car, falls down a hoistway, being caught between an elevator car and a hoistway wall or counterweight, and being struck by the closing hoistway doors.

Fortunately, in nearly all stalled-elevator incidents passengers are safe if they remain inside the elevator car. It is when they become impatient, panicky, or simply curious that they sometimes attempt to leave the elevator car before qualified help arrives. It is during or after this time that passengers often succumb to the dangers of falling down the hoistway.

Even members of the elevator industry are sometimes injured or killed when working in or around elevators. Firefighters can become victims as well. Carelessness, complacency, and ignorance are human factors that often contribute to tragedy.

Most stalled-elevator incidents are passenger inconveniences, not emergencies. Yet, during passenger removal operations, firefighters must always be aware of the potential risk of harm to passengers as well as themselves. The most important aspect of any elevator incident is the safety of passengers, victims, and firefighters.

The safety principles discussed later in the chapter also apply during practical training sessions. In any case, firefighters must remember that an incident involving a normally safe elevator can quickly turn deadly if safety principles are ignored. Firefighters' assignments at the scene of an elevator incident must be commensurate with their level of training and experience.

Elevator Hazards

Part of being safe is knowing the hazards. Even though the elevator is a safe means of public transportation, it can pose potential hazards to firefighters during training and elevator rescue operations. The major concerns are the fall hazard and the unexpected movement of an elevator car.

A fall from heights can originate at an unprotected landing (floor) or from top of an elevator car. Even when a hoistway door is in its closed and locked position, firefighters must remember that the hoistway is just on the other side. The hoistway door is attached at the top to the hoistway wall and also has gibs attached to the bottom of the door that slide in a metal groove. The gibs slide in the track as the hoistway door opens and closes. Over time, the gibs can wear and make the door loose in the track.

If firefighters attempt to force open a hoistway door by breaking the door gibs or otherwise lifting the door from its track, they just might end up at the bottom of the hoistway along with the hoistway door! We should know that the hoistway door-locking mechanism for the hoistway door is mounted at or near the top of the hoistway door. This is where firefighters should apply force with a forcible entry tool to open a hoistway door, not at the bottom of the door. For this reason, firefighters should avoid pushing or leaning against any hoistway door. You don't know the condition of the door. Be mindful that the hoistway door is what stands between you and the deadly hoistway.

We provided this information to alert firefighters that the danger of a hoistway is ever present, regardless of whether you see the hoistway. The following safety principles include how to guard an unprotected hoistway.

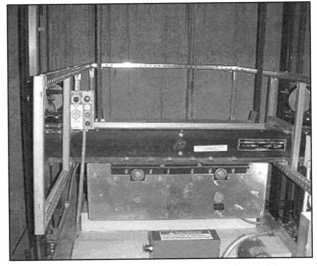

Fig. 11–1. A permanent barrier installed on three sides of car to reduce fall risk to the elevator mechanic or inspector (and firefighters)

Firefighters must also be aware of the fall hazard when working atop an elevator car. Hoistway walls surround the elevator car. Usually, there is no protective barrier around the car top to protect firefighters. However, in modern elevator installations, a barricade is often permanently installed to keep elevator mechanics and inspectors away from the exposed hoistway (fig. 11–1).

Other Hazards

A hazard control sequence has six elements.

1. *Anticipate* hazards through training and field experience.

2. *Identify* hazards visually on scene.

3. *Evaluate* the hazards. Do they pose imminent danger?

4. *Alert* firefighters of the hazards.

5. *Mitigate* the hazards as is reasonably achievable.

6. *Monitor* the hazards to prevent firefighters from getting too close to them.

When the elevator incident requires firefighters to work from the top of an elevator car, check from the entry landing for any hazards *before* getting onto the car.

The survey can reveal

- Tripping hazards

- Limited working space

- Slippery surface

- Unprotected hoistway (fig. 11–2)

- Equipment attached to hoistway walls

- Moving parts (car door operator, 2:1 roping and counterweight; figs. 11–3 11–4, and 11–5)

- Lightbulb without guard

- Damaged electric wiring

Fig. 11–3. Be aware of the elevator car's door operator assembly's moving parts.

Fig. 11–4. Two-to-one roping. When the elevator car moves, so do the hoist ropes around the sheave. This is another reason to ensure that the mainline power to the elevator is shut off before getting on top an elevator car.

Fig. 11–2. An unprotected hoistway can lead to a fatal fall for an unsuspecting firefighter or building occupant.

Fig. 11–5. Another example of 2:1 roping

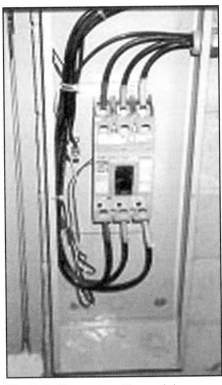

Fig. 11–6. Shunt trip with panel door open

If firefighters follow the safety principles and rules discussed later, their risk of harm from these hazards will be minimized. Unfortunately, some firefighters might fail to heed our warnings and attempt to do rescue operations without removing mainline power from elevators and not using required personal protective equipment. We all know what the consequences can be—injury or death.

Shunt Trip and Door Restrictors

Although both of these devices are discussed in detail elsewhere in the book, we believe that also it is worth mentioning them here. The shunt trip is installed when the elevators are protected by an automatic fire sprinkler system (fig. 11–6).

Given the purpose of the shunt trip, it is a source of controversy among firefighters. When a fire in the machine room or hoistway actuates a heat detector, the shunt trip later actuates to remove all power to the elevators before water flows from a sprinkler head. If firefighters were unfortunate enough to be inside one of the elevators when the shunt trip actuated, the loss of power would strand or maroon them inside the ill-fated elevator car.

We recommend that you determine if shunt trip is installed in buildings equipped with an automatic fire sprinkler system. In those cases where they are installed, you should identify ways to safely escape the elevator car by using forcible entry tools or await rescue by other firefighters.

The door restrictor is provided on elevators to prevent passengers from opening the car door when the elevator car is located between 3 and 18 inches above and below a landing (fig. 11–7).

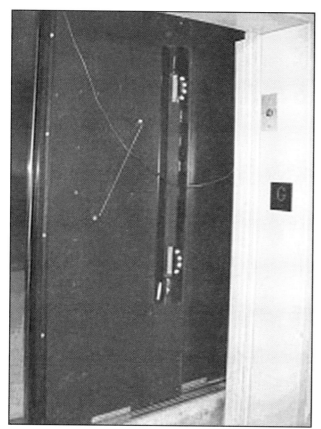

Fig. 11–7. A type of door restrictor

From a human perspective, three distracters of safety in the fire service are complacency, carelessness, and ignorance. These robbers of safe practices not only impact firefighters but also elevator constructors, mechanics, helpers, and inspectors. For example: "Ah, we've done this a dozen times! There's no need to kill the power." It is when firefighters develop a "no big deal" attitude (complacency) that one of the risk flags is raised. Now, consider carelessness. It's things we do without giving the necessary and proper time and attention. "Oops! Sorry about that, would you send that tool back up here?" And the third distracter, ignorance: "Oh, I didn't know that. I forgot to read the revised SOG (standard operating guideline)."

We well know that firefighters would not do something intentionally that would either harm themselves or others. Avoiding the CCI syndrome (complacency, carelessness, and ignorance), firefighters can continue doing safe operations. In this chapter we share our philosophy in more detail. Also, we present a set of principles for consideration by firefighters who have yet to develop standard operation guidelines or are in the process of reviewing them for possible need for revision.

Prevent elevator car movement

Removing power from the stalled elevator (and other elevators as needed) is crucial to the safety of both firefighters and passengers (fig. 11–8). Applying the stalled elevator's emergency stop switch, if available, also is recommended.

Lockout/tagout is a safety measure used in conjunction with removing power from a stalled elevator. It is intended to prevent injury and death to firefighters. We recommend that firefighters carry lockout/tagout equipment on their emergency vehicles. Although figure 11–9 shows locks in the machine room, don't expect to find them in most machine rooms. The Occupational Safety and Health Administration (OSHA) Rule 29 CFR 1910.147 on the control of hazardous energy sources is used by firefighters to prepare elevator safety guidelines. Highlights of the OSHA lockout/tagout standard are provided later in this chapter.

Before the advent of restrictors, it was easy for passengers to open a car door just about anywhere along travel of the elevator car. This was done in haste to escape the stalled elevator car or in the interest of mischievousness. A frequent outcome, in either case, is one where the passenger either is injured or killed because of tripping or a fall.

Safety Philosophy

Generally, firefighters who have good attitudes about safety turn those attitudes into safer elevator practices. We know that during training and at a routine or emergency elevator incident, attention to safety usually is not where the action is. Yet, it's where a major focus must be to develop or reinforce the right safety attitude, in order to promote safer practices that serve to reduce the risk of injury.

Fig. 11–8. Mainline power disconnect, one for each elevator (1 and 2)

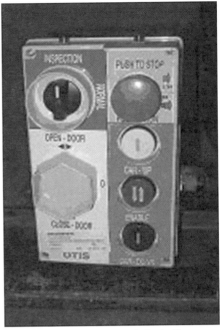

Fig. 11–10. Car top controls including inspection and emergency stop switches

Fig. 11–9. Locks available in machine room for use in lockout/tagout operations

In addition to shutting off the power to the elevator and using lockout/tagout procedures, firefighters should set the emergency stop switch located inside the stalled and rescue-assist elevators, if provided, to the *safe mode* position. We also recommend that the emergency stop and inspection switches located on top of the elevator car be set in the *safe mode* position before anyone steps onto or works from the car top (fig. 11–10).

If elevator equipment is damaged, and there is concern about whether the elevator car still could move, firefighters should consult with the on-scene elevator mechanic to obtain advice as to how best to secure the elevator car.

Guard open hoistway

An exposed, unprotected hoistway can threaten the lives of unsuspecting passengers and firefighters. Firefighters already must be in position, with a full body harness and secured lifeline and ready at the hoistway doors before the doors are opened, exposing the hoistway. Placing a short ladder across the hoistway opening is one way to barricade the unprotected open space. Firefighters should secure the ladder to prevent its movement.

If the nature and scope of an emergency elevator incident require that the hoistway door be kept open for an extended period, firefighters should consider using a door wedge tool. One type of tool is shown in figure 11–11. Unlike the nonrecommended use of a screwdriver, clothespin, or a wooden or rubber doorstop, the door wedge tool is specifically designed for use to hold open a hoistway door. Elevator hoistway doors are required by the elevator code to close automatically when the car is not present at the landing. Although the door wedge tool is designed to hold open hoistway doors, it still is necessary for the firefighter to continue to stand at the open hoistway door until it is no longer needed and the hoistway door is returned to its fully closed and locked position.

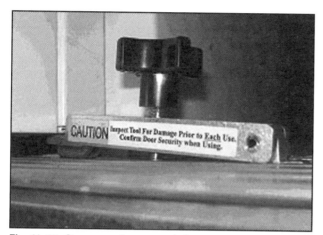

Fig. 11–11. One type of commercially available door wedge tool

Protect passengers

As we already know, unless passengers are endangered by fire, smoke, or other dangerous situations, they are safe remaining in the elevator car. However, it is during passenger-removal operations that the risk of injury can dramatically increase, unless appropriate safeguards are taken. They could "slip, trip, or fall" unless firefighters take strict measures to minimize the risk. If the elevator car is located near a landing, and firefighters can remove the passengers through the elevator car entrance, firefighters need only reassure passengers and maintain a handhold on them just before, during, and after removal to a landing. However, don't attempt to remove passengers through the entrance until they are properly secured and the space below the elevator car is barricaded. The open hoistway has been the cause of many elevator fatalities, as one's body tends to arch under the car when exiting.

Additionally, if the situation requires that firefighters must remove a passenger by the elevator car's top emergency exit and then to a landing, they should place a helmet on the passenger's head and secure the strap to prevent helmet loss, use a safety harness and rope system, and maintain a handhold on the passenger as he or she is removed from the elevator car until they are assisted by other firefighters at the landing. If power tools are required to help with removal of a pinned person, firefighters should implement measures such as having the passenger wear safety glasses or goggles, or covering the passenger with a firefighter's turnout coat.

Protect firefighters

The risk of possibly falling down a hoistway and being struck by an object are two major concerns of firefighters. For these reasons, firefighters must wear head, eye, foot, and hand protection, and use fall-protection equipment.

Another important safety measure is to assign a safety officer. This person should be a member of the elevator rescue team, preferably a firefighter who is experienced in elevator emergency management and is safety minded.

When developing or revising elevator rescue guidelines, be sure to sufficiently address safety measures. Providing good training on elevator rescue procedures and proper use of tools and equipment promotes greater safety awareness and provides for safe practices by firefighters.

The Elevator Safety Team

The incident commander, firefighters, safety officer, and elevator mechanic are part of the elevator safety team. Through their attention to adhering to safety principles and good practices, the risk of injury during training or incident management should be minimal.

Elevator mechanics are an important part of elevator safety. They are among the experts in the elevator industry. They know how the elevators work, what causes them to stall, and how to make them run again.

Whether training or managing an elevator incident, the elevator mechanic is a valuable resource to firefighters. Having an elevator mechanic on the scene, especially if the situation will require firefighters to work near or in the hoistway, makes for safer, more efficient rescue operations.

Safety Rules

The following four rules embrace our safety philosophy and principles. The rules are modified here to represent a generic format for consideration and use by other fire departments.

Rule 1

- Whenever possible, leave the occupants in the car until an elevator mechanic arrives.

- They are only "locked in a box" and are safe until we arrive and endanger them.

- The typical service contract has a two-hour response time.

- Be patient!

Rule 2

- If the passenger is not in danger, notify the elevator company, if it has not been called already. Wait for the elevator mechanic to arrive, then ask the mechanic to move the car to a landing and open the doors to allow passengers to exit the car in a normal fashion.

- *Before* any attempt is made to remove a passenger from an elevator, power *must* be disconnected and lockout/tagout procedures followed at the main power disconnect switch/panel located in the machine room.

- This rule is not debatable and must be strictly *enforced*.

- This procedure is referred to as *power down* and must be documented.

Rule 3

- Never move an elevator.

- Only an elevator mechanic should attempt that procedure.

- The elevator mechanic who showed you how to do it will testify against you in court, because it typically is against elevator

company policy to show others how to move an elevator by other than the normal operating buttons.

- They will deny they ever trained you to do it. Ask them for a letter on company stationery requesting that you do it.

- You will *never* get it!

Rule 4

- The response must be recorded.

- Refer to local standard operating guidelines (SOGs) for firefighter service problems.

- Follow local SOGs for entrapments and removal of passengers.

- Lockout/tagout must be in writing.

These rules are based on field experience. The common theme expressed is one of safety.

Lockout/Tagout (OSHA CFR 29 Part 1910.147)

Working near energized electric equipment without first removing its main power lends itself to two potential consequences for firefighters. The first is electrocution from contact with exposed electric components. The other is the sudden, unexpected movement of the elevator or other movable equipment that can injure or kill firefighters. To eliminate these potentially deadly consequences, firefighters must not only remove the main power to a stalled elevator and other elevators as necessary but also implement added measures to prevent restoring of power to the elevator(s).

The OSHA standard "Control of Hazardous Energy (Lockout/Tagout)" is found in Title 29 of the Code of Federal Regulations (CFR) part 1910.147. This standard details the steps employees must take to prevent accidents associated with hazardous energy. Moreover, it addresses practices necessary to disable machinery and prevent the release of potentially hazardous energy while maintenance or servicing activities are performed.

Four definitions included in the OSHA lockout/tagout standard are provided here for firefighters who are not familiar with lockout/tagout:

Lockout. The placement of a lockout device on an energy-isolating device, in accordance with an established procedure, ensuring that the energy-isolating device and the equipment being controlled cannot be operated until the lockout device is removed.

Lockout device. A device that uses a positive means such as a lock, either key or combination type, to hold an energy-isolating device in the safe position and prevent the energizing of a machine or equipment. Included are blank flanges and bolted slip blinds.

Tagout. The placement of a tagout device on an energy-isolating device, in accordance with an established procedure, to indicate that the energy-isolating device and the equipment being controlled may not be operated until the tagout device is removed.

Tagout device. A prominent warning device, such as a tag and a means of attachment, which can be securely fastened to an energy-isolating device in accordance with an established procedure, to indicate that the energy-isolating device and the equipment being controlled may not be operated until the tagout device is removed.

Many firefighters already have written lockout/tagout procedures consistent with the OSHA lockout/tagout standard. However, firefighters who do not have written lockout/tagout procedures are cautioned to prepare, train on, and start using their procedures as soon as possible. It is an OSHA requirement. Checking with neighboring departments that already have OSHA-compliant lockout/tagout guidelines can serve as a model for others to use. Also, using similar guidelines among neighboring fire departments will help facilitate their use and help eliminate any misunderstanding.

Safety Highlights and Considerations

- Power down elevators and use lockout/tagout procedures.

- Assign firefighters to guard entrance to any unprotected hoistway.

- Use a barricade such as a short ladder to block hoistway opening as required.

- Wear head, eye, hand, and foot protection, and use fall-protection equipment.

- Use ladders to assist passengers from a stalled elevator through the top emergency exit or normal car entrance.

- Do not use a side emergency exit. The ASME 17.1 Code no longer recognizes it as a safe means of passenger removal and prohibits it use (fig. 11–12).

Fig. 11–12. The ASME 17.1 Code prohibits use of the side emergency exit.

- Do not push or lean against any hoistway door.

- Do not attempt to force open a hoistway door by applying force near bottom part of the door.

- Check the top of the elevator car for hazards from a landing before stepping onto the car (figs. 11–13 and 11–14).

Fig. 11–14. The counterweight moves silently in the hoistway and can be deadly unless the mainline power is shut off.

Fig. 11–13. The car-top smooth surface, car-door-operator assembly, and open hoistway are common hazards found on top of an elevator car.

- Remember the hazard control sequence: anticipate, identify, evaluate, alert, mitigate, and monitor the hazards.

- Assign a safety officer to monitor hazards and address any unsafe practices.

- Use the expertise and resourcefulness of an elevator mechanic.

- Coordinate activities between the incident commander and the elevator mechanic.

- Note the members of the elevator safety team: incident commander, firefighters, safety officer, and elevator mechanic.

- Document the elevator incident. If the incident involves injury or a fatality, use photographs and diagrams, and take witness statements. Notify the appropriate authorities.

- Review significant incidents with members who were on the scene as well as others to gain insight; identify things that went right and things requiring further review or change.

- Be sure there is sufficient training to ensure that firefighters can safely use elevator rescue tools and equipment.

- Remember, if you should observe slack in hoist ropes in a machine room or on top of an elevator car, the elevator car can be in danger of falling. Notify the elevator mechanic.

Summary

The safety of firefighters and passengers is and must continue to be of paramount concern. Though a safe means of public transportation, if an elevator stalls with passengers aboard who later attempt to leave before qualified help arrives, or a person gains access to the top of an elevator to foolishly jump to a passing elevator or counterweight ("elevator surfing"), injury and death can result. In almost all cases, passengers inside a stalled elevator are safe as long as they remain there until arrival of an elevator mechanic, qualified firefighters, or a qualified building engineer.

Regardless of whether an elevator incident is a nonemergency or emergency, firefighters must ensure that their operating guidelines give appropriate attention to safety. Preventing movement of an elevator, guarding open hoistways, and protecting both firefighters and passengers are crucial to any and all elevator incidents.

The four rules listed in this chapter are based on field experience and are provided as a guide for other firefighters to consider when developing or revising their own elevator rescue guidelines. Removing power from elevators, using lockout/tagout procedures, and thoroughly documenting each elevator incident are examples of things firefighters must address. Firefighters who already have and use written lockout/tagout guidelines are better prepared than firefighters who have yet to prepare them. To be without lockout/tagout guidelines can result in injury and possibly civil action.

The more firefighters practice the safety philosophy by adhering to the four safety rules or similar guidelines, the safer firefighters will be while engaged in elevator rescue operations. Also remember who is part of the elevator safety team: incident commander, firefighters, safety officer, and elevator mechanic. Stay safe.

Review Questions

1. List and describe the four safety rules discussed in the chapter.

2. Who makes up the elevator safety team?

3. What are the three human factors discussed in the chapter that often contribute to tragedy?

4. List six hazards that firefighters may find on or near the top of an elevator car.

5. Explain what is meant by the terms lockout and tagout.

6. List the five items of personal protective equipment firefighters should wear or use during rescue operations.

Field Exercise

Conduct or request an elevator safety and survival training session to review and discuss the safety principles, rules, and other safety points noted in the chapter.

Chapter 12
Rescue Tools and Equipment

Introduction

Successful elevator rescue operations depend, in part, on having the right tools and equipment available at the scene. Firefighters open most hoistway doors by using a hoistway door unlocking device key, an elevator pole, or, as a last resort, a forcible entry tool.

As important as having the right tools and equipment is having firefighters available who are trained or experienced in their proper use. Trial and error is the least preferred method of acquiring skill level, especially if a mistake results in injury. An example is a firefighter who does not know how to safely use a hoistway door unlocking device key. After unlocking the hoistway door and while still holding onto the key, the firefighter quickly moves the door to its full open position. Unless the firefighter lets go of the key before it hits the doorjamb, serious hand injury could result. Although the collar provided on some types of hoistway door unlocking device keys, is designed, in part, to prevent injury, improper holding of the key could contribute to injury upon unfortunate contact with the doorjamb! By removing the hoistway door unlocking device key from the hoistway door after moving the keeper hook and sliding the door clear of the interlock, the key is removed, and the door then is safe to open fully.

Tools are of little or no value if there is no one available who knows how to properly use them. It's not so much the forcible entry tools that firefighters have trouble with but rather tools specifically designed for elevators. Again, good training affords firefighters the opportunity to gain proficiency in the safe and proper use of tools and equipment.

Included is a list of items that firefighters should find useful during elevator rescue operations. Most elevator incidents are relatively simple to manage because they involve passenger entrapment without injury or illness. As you should recall, passengers who remain inside the elevator car are safe. It is when they try to leave before help arrives, or even during rescue operations by firefighters, that they can be injured or even killed. Getting outside the "box" can be dangerous, unless the right safeguards are followed.

With the exception of items such as forcible entry tools, ladders, lights, and other tools and equipment that have universal fire service application, firefighters should conveniently group elevator tools together on emergency vehicles.

A basic set of elevator rescue tools and equipment includes hoistway door unlocking device keys, elevator-poling tool, firefighters' service key, forcible entry tools, portable radios, lockout/tagout kit, door wedge tool, hand lights, and lifeline rope and harness. Initial information provided by the emergency communications center (ECC) can alert firefighters to the need for additional equipment. An example would be a call for a reported person pinned between elevator equipment. This is no ordinary incident!

Lockout/Tagout Equipment

Using lockout/tagout equipment is an "insurance card" to prevent unauthorized and premature restoring of mainline power to the stalled elevator and adjacent elevators as needed. Don't leave the station without it!

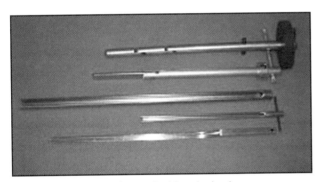

Fig. 12–1. Hoistway door unlocking device keys

We recommend that you carry an assortment of hoistway door unlocking device keys (fig. 12–1). Surveying your elevator buildings can reveal the need to carry other types of keys as well. Even though these keys might be available in a security box or Knox-Box mounted near the elevator entrance, we still recommend that you carry your own set of keys on emergency vehicles such as your technical rescue response vehicle, ladder truck, or rescue squad. What do you do if you respond into another district's area for one or more stalled elevator incidents because of a power outage? Will the hoistway door unlocking device key be available on site? A basic set of hoistway door unlocking device keys include the following:

- Single-link drop key
- Double-link drop key
- Large and small lunar (half-moon) keys
- Tee key

Firefighters' service key

The firefighters' service key is used to operate Phase I and Phase II operations of firefighters' service. Some jurisdictions have an ordinance that requires the use of a standard firefighters' service key for elevators within their local response area. Actuating Phase I operations might override the problem and recall the elevator to the lobby level to allow passengers to exit the car. This tactic will only work as long as the safety circuit has not been opened.

Elevator-poling tool

There are different tools used to reach and unlock hoistway doors. A pike pole is sometimes used to pole down and pole across. However, in the latter case, the pole may not be narrow or flexible enough to freely move through the space between the front of an elevator car and hoistway wall to reach the hoistway door unlocking device.

Some firefighters use a thin, flat pole or even a long wooden handle as a poling tool. The best way to determine the benefit and limitations of a poling tool is to set up a training session with an elevator mechanic and try it (fig. 12–2).

Fig. 12–2. A common tool used to unlock a hoistway door

Hand lights

Frequently these lights are used to help firefighters locate a stalled elevator car and a hoistway door's interlock release roller during poling operations, identify potential hazards in the hoistway and on top of an elevator car, and improve lighting in dimly lighted areas. Firefighters should consider carrying a section of rope with which to attach and lower a portable light down the hoistway as needed.

Rescue rope and harness

Firefighters may need rescue rope and harnesses to help ensure the safety of passengers when removing them through the elevator car's top emergency exit. Be sure to use lifelines, not just any rope. It is also important that the lifelines are properly applied by a firefighter to the fall-arrest harness before allowing the passenger onto the car top.

Forcible entry tools

Although emergency elevator incidents are seldom encountered, firefighters must take into account this possibility. Forcible entry may become a necessity. The selection of forcible entry tools is a matter of preference. However, the more utility it has in the area of prying and breaching, the more practical it should be.

Chapter 16 covers a selection of forcible entry tools. These are commonly used by firefighters in day-to-day responses. There are more. What is important here is having the right tools available for elevator extrication incidents. Set up a meeting with a building's elevator service mechanic through the building owner or manager to discuss ways to lift or support an elevator car, or how to obtain lateral movement of the elevator car to free a victim. Through this discussion, firefighters can identify different methods and tools required to accomplish these important tasks.

EMS jump bag

The EMS jump bag is a common item carried on EMS vehicles. It is not needed on most elevator incidents. However, when there is a pinned person or a person who has fallen down an elevator shaft, the EMS jump bag (at a minimum) is needed to attend to any victims. Part of the ECC's role in an extended incident is to gather the resources that we will need. If we do not tell them exactly what we want, then we will not get it. Tell them, and they will get the world for you.

Portable radios

In many fire departments each firefighter is assigned his or her own portable radio. During an elevator rescue incident, communication is needed among firefighters assigned to power down and lockout/tagout and firefighters directly involved in the rescue operation. If it is necessary for the elevator mechanic to use the elevator's controller to move the elevator car or use other means to lower or raise the elevator car, then the elevator mechanic should have a radio, also. This would allow the elevator mechanic to communicate and coordinate operations with the incident commander. Since a firefighter is assigned to guard entry to the machine room, that firefighter's radio can be shared with the elevator mechanic.

The radios are also used to communicate with mobile units and the ECC to provide progress reports and to request additional specialized resources such as a hospital's trauma team.

Door wedge tool

The door wedge tool is used by elevator mechanics to hold open a hoistway door during maintenance and repair of an elevator (fig. 12–3). It is a much safer way than using a clothespin or screwdriver. However, when using the door wedge tool, be sure that the hoistway opening is still guarded by a firefighter and a barricade is used if conditions allow. The door wedge tool is one of several available. The one shown is for illustration purposes only. It is not an endorsement.

Fig. 12–3. One type of door wedge tool

Elevator kit

Using a toolbox to carry small tools such as hoistway door unlocking device keys, flashlights, lockout/tagout equipment, door wedge tool, screwdrivers, and wrenches is a convenient way to keep them together for quick access at the scene of an elevator incident and later for inventory.

Ladders

The elevator incident may require the use of one or more ladders to reach top of the elevator car, to aid in removing passengers from inside the car by either using the normal car entrance or the top emergency exit, or to barricade an unprotected hoistway opening.

Flood lights

Lights are often useful during extended rescue operations or when working under poor lighting conditions. Special-call a company specifically to run and maintain your lights if possible.

Fall-Protection Equipment

Fall-protection items are needed to ensure safety of firefighters who are required to work near or inside the hoistway or from top of an elevator car. This equipment helps to prevent firefighters from becoming victims of a fall from heights.

Other Forcible Entry Tools

Based on the type of incident and existing conditions, firefighters may need additional equipment. The type and general construction of the hoistway door and walls, the type and location of the interlock, and the construction and operation of the car door are factors that influence which forcible entry tool is best suited for the rescue operation.

Summary

Probably 50% of successful elevator rescue operations is having the right tools available for the right type of incident. Most of the stalled elevator incidents with passenger entrapment can be managed using a basic set of elevator rescue tools and equipment. Yet, it is also important that firefighters receive initial and refresher training in the safe and proper use of these tools and equipment.

The more complex an elevator incident, the more likely is the need for additional equipment. Identifying and having ready tools and equipment to manage different elevator incidents can help streamline rescue operations and minimize risk of injury.

Review Questions

1. During elevator incidents, most hoistway doors are opened by firefighters using what three methods?

2. To facilitate elevator rescue operations, firefighters should have a basic set of elevator tools and equipment with them when they enter the building. List eight items.

Field Exercise

Using the basic set of elevator tools and equipment noted in this chapter as a guide, check the equipment inventory of your emergency vehicles that typically respond to elevator incidents for these items. Note any items that are not available, and decide which ones to acquire. Grouping items together on the emergency vehicle(s), where practical, can facilitate access and routine checking.

Chapter 13
Elevator Hoistway Door Unlocking Device Keys

Introduction

In the elevator industry, handheld devices specifically designed to unlock hoistway doors are referred to as hoistway door unlocking device keys, interlock door release keys, and escutcheon keys (fig. 13–1).

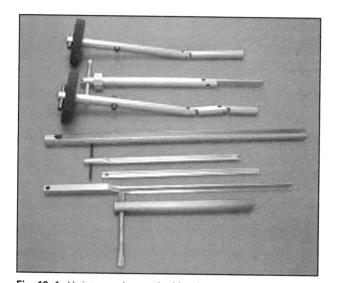

Fig. 13–1. Hoistway door unlocking keys

The safe use of hoistway door unlocking device keys requires a good understanding of how an elevator hoistway door opens, closes, and locks. Moreover, practical training sessions are needed to help firefighters gain both competence and confidence in the proper and safe use of hoistway door unlocking device keys. The chapter discusses the function of the interlock, types of keyholes and escutcheons, and different types and operations of hoistway door unlocking device keys.

Elevator mechanics and inspectors use these keys to gain access to the hoistway of elevators to do inspections, maintenance, and repairs. Firefighters use hoistway door unlocking device keys to open hoistway doors during training and actual elevator incidents. Today, in most jurisdictions with elevator buildings, firefighters usually carry hoistway door unlocking device keys on emergency response vehicles.

Most of the hoistway doors unlocking device keys discussed in the chapter are commonly found in buildings with modern elevator installations. The ASME A17.1 Code allows them for use on new elevators. Other types of hoistway door unlocking device keys are still in use for older elevator installations. Yet, the basic principles of their operation are the same or similar to the keys in use today.

Firefighters should become familiar with the types of hoistway door unlocking device keys used in both old and modern elevator installations located in their response area. Regardless of the age of elevators and the types of keys required, safety is the paramount concern during training and actual use of hoistway door unlocking device keys. However, very old elevator installations are not likely to have updated safety features found on modern elevators. The local elevator mechanic or inspector is an excellent source of information on both old and modern elevator installations. Prudent firefighters would welcome their expertise and advice in the interest of safety and operational proficiency.

Interlock

Before discussing the types and operations of hoistway door unlocking device keys, it is important to explain the interlock and its twofold purpose. This electrical and mechanical safety feature is usually mounted inside and above the hoistway door on the hoistway header beam. It has a dual function: First is to electrically open and close a contact to either prevent or allow movement of the elevator car. Second, the interlock serves to mechanically lock the hoistway door when the door is fully closed and also unlock the door to allow it to move to its open position.

On a common type of interlock the keeper mechanically locks the hoistway door when the keeper hook is in the closed position (fig. 13–2).

The keeper hook is attached to the hoistway door release roller usually through a lift rod. When the release roller is moved to unlock the hoistway door by action of the elevator car door's clutch assembly or movement of the door unlocking mechanism by using a hoistway door unlocking device key, the keeper hook is moved away from the keeper (fig. 13–3). Another type of interlock is shown in figures 13–4 and 13–5.

Fig. 13–3. Interlock with keeper hook in open position (courtesy of Jim Jarboe)

Fig. 13–4. Another type of interlock with keeper hook in closed position (courtesy of Jim Jarboe)

Fig. 13–2. Interlock with keeper hook in closed position

Fig. 13–5. Interlock with keeper hook in open position

The key unlocking bolt replaced an earlier design known as the adjustable key unlocking clamp or "flag." The keeper hook also has an electric bridge that makes up a contact. This is a feature known as the *safety circuit*. When opened, it causes the brake to be applied on the drive machine in the machine room. When the keeper hook is in the closed position, both electric contacts (one mounted inside the interlock box and the other on the keeper hook) are closed, thus completing the hoistway door electric circuit. Then, as far as that hoistway door is concerned, the elevator car is free to travel to other landings.

Conversely, when the contacts are moved apart or opened when moving the keeper hook, power to the elevator is removed from the brake of an electric traction elevator or hydraulic pump of a hydraulic elevator, preventing the car from moving. The interlock box usually is covered with a metal plate that serves to protect the interlock components and prevent accumulation of dust. However, a word of caution: When an interlock is open, power can and is applied to the driving machine for leveling. This means that the elevator still can move if it is near a landing to bring the car level with the landing.

Although there are different types of hoistway door interlocks, they basically operate in a similar fashion. Firefighters should seek assistance from an elevator mechanic, inspector, or member of the fire service with known expertise in the subject of elevator safety and operation.

Hoistway Door Keyhole

A hoistway door keyhole is provided on many hoistway doors. It is usually located near the top of the door. However, it may also be found lower down from the top of the hoistway door. The keyhole is either drilled through or cut into the hoistway door, depending on the type of keyhole. In some cases, a metal sleeve is installed through the keyhole to facilitate use of particular types of hoistway door unlocking device keys.

The keyhole provides an opening for use of a hoistway door unlocking device key, commonly referred to in the fire service as a hoistway door key. The keyhole is usually round in shape. However, another type of keyhole cut into a hoistway door is the small lunar or crescent-shaped hole. This type is also referred to as half-moon. Another shape is the T shape.

Escutcheon Plate

An escutcheon is a protective plate used to cover the keyhole. The cover is designed to facilitate use of a particular type of hoistway door unlocking device key. It also helps to prevent unauthorized persons from unlocking the hoistway door by using a device other than the required hoistway door unlocking device key. Common types of escutcheon plates are round, large lunar, and inverted tee (figs. 13–6, 13–7, and 13–8).

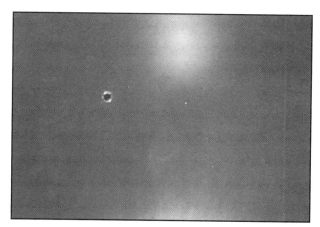

Fig. 13–6. Round

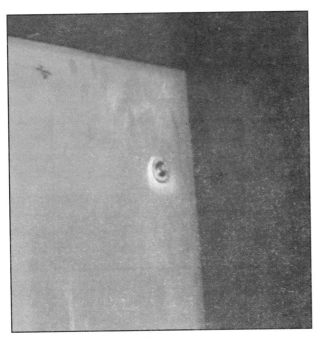

Fig. 13–7. Lunar (half-moon)

Fig. 13–8. Tee

However, not all hoistway doors are provided with hoistway door keyholes. For example, the elevator safety code adopted by California does not permit the use of hoistway door unlocking devices. Therefore, no keyholes are provided either in the hoistway door or on an adjacent wall. This action was motivated by the concern for public safety and the potential for liability that can result from an elevator incident involving injury or death. Unauthorized access to an elevator's hoistway can lead to damage to elevator equipment, personal injury, and even death. Until recently, the elevator safety code in Massachusetts also prohibited the use of hoistway door unlocking device keys.

While elevator safety codes in use by most states do allow the use of hoistway door unlocking device keys, they are not necessarily provided on all hoistway doors. Before 1996, the ASME 17.1 Code required, in effect, that hoistway door keyholes be provided on the lowest and an upper landing served by an elevator. The upper landing was often the top floor. Since 1997, the ASME 17.1 Code has required that a keyhole be provided on hoistway doors on every landing served by an elevator. This means, consistent with the ASME 17.1 Code (1997 and later) and where adopted by local or state jurisdictions, firefighters have access to keyholes for use of hoistway door unlocking device keys.

Hoistway Door Unlocking Device Keys

Before discussing the types and operation of these keys, it is important to note that training in their use should occur at the lowest landing where the hoistway door keyhole is provided. If this cannot be done, then position the elevator car at the landing immediately below the landing where training will occur. This is the safest way because some pits are quite deep. If the fall hazard with the door open is more than 6 feet, fall protection should be worn per OSHA regulations.

After a firefighter calls for and gets into the empty elevator car affected by the training session, the firefighter should take the car to the desired landing and remain until the mainline power disconnect to that elevator is in the off position and the lockout/tagout procedures are followed. If present, apply the emergency stop switch.

There are many types of hoistway door unlocking device keys used to unlock hoistway doors. The most commonly found keys in use today are the drop key, small and large lunar keys, and the tee key. With the exception of the small lunar key, escutcheon covers are usually found on hoistway doors accessible by the drop, large lunar, and tee hoistway door unlocking device keys.

A hoistway door unlocking device key is a valuable tool for use by firefighters. Yet, the use of a hoistway door key carries with it many safety precautions and responsibilities for firefighters. Just consider that when the key is used to unlock a hoistway door and the door is then moved to the full open position, the risk of harm to firefighters increases substantially. An unprotected hoistway is a potentially deadly situation. Safety must continue to demand appropriate attention.

As discussed in the chapter 11, Elevator Safety, there are four general safety principles that firefighters must address before, during, and after management of an elevator incident. These include preventing the elevator car from moving; guarding the hoistway opening before, during, and after a hoistway door is opened; and protecting firefighters and passengers.

Hoistway Door Unlocking Device Key Box

Although the ASME 17.1 Code has prohibited for many years the use of a special box to hold a hoistway door unlocking device key, they are still found (figs. 13–9 and 13–10). When they are present, firefighters should request that the building owner or manager replace the box with a security-designed box. This type of box is used to hold certain building access keys that are often used by firefighters when they respond to an emergency. That box or a larger model can hold the hoistway door unlocking device key. Generally, the key required to unlock these boxes has a unique lockset that requires use of a special-cut key designed to prevent duplication.

Fig. 13–9. Key box

Fig. 13–10. A key box with a hoistway door unlocking device key. A rare sight!

Usually, for security and convenience reasons, one key fits all boxes, and all keys are under sole control by firefighters in that response area. The supplier of the keys and boxes provides them directly to the fire department. A written authorization agreement between the supplier and the fire department is required.

In some jurisdictions the key is also made available to neighboring fire and rescue departments that provide mutual aid response to that fire department. However, the host department usually must give authorization in writing both to the supplier of the keys and the neighboring fire departments. Although this key can be beneficial during emergency operations, the loss of such a key could cause both frustration and concern to firefighters. Security in the storing of these keys is very important and should require appropriate measures to minimize theft or loss. In buildings with good security and maintenance programs, the key is kept in another location accessible to the building engineer, manager, or firefighters.

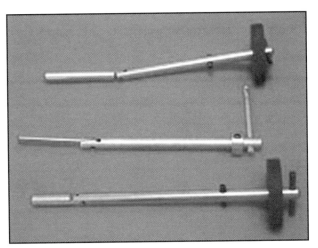

Fig. 13–11. Three types of single-leaf drop keys

The drop key is one of the most commonly used hoistway door unlocking device keys. Some of the other names used to describe this type of key are toggle, flip over, broken arm, and flapper (fig. 13–11).

The drop key has three basic components: tee handle, cylindrical shaft, and a hinged flat or round link (leaf). These links are also referred to as pins or bars and are found in different lengths. Some drop keys have a second or third hinged link. Others have adjustable stop collars.

The double or triple link is required where there is inadequate clearance between the hoistway and car doors to permit use of a single-link drop key. As a result, the link cannot drop to its required vertical position on the hoistway side of the door. The operation of the multiple-link drop key is discussed in more detail later in the chapter.

The adjustable stop collar is for safety of the user. If the door powers open when the key is in the hole, the collar protects the key user from pinching fingers on the doorjamb. A better way to avoid injury is to remove the key from the hoistway door as soon as the key mechanically unlocks the door and the doorkeeper hook clears the interlock box.

After using a small Allen wrench or similar tool to loosen the collar's setscrew, the collar is free to slide along the shaft of the drop key. Once the key is positioned to permit the link to drop on the other side of the hoistway door and is operated to ensure that the link is the right distance from the hoistway door, the setscrew is tightened. Now, whenever the drop key is needed to unlock that hoistway door, it simply is aligned with the keyhole and inserted into the stop collar. The link is now at the proper distance from the hoistway door to engage the door release. Of course, this collar adjustment is convenient for drop keys that are maintained on a building's premises and where clearances are the same.

However, the drop keys carried by firefighters should have the adjustable stop collar positioned and tightened near the tee handle of the key (fig. 13–12). This will allow the key to be used on hoistway doors that require different clearances to properly align the link on the hoistway side of the door.

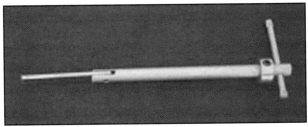

Fig. 13–12. Adjustable stop collar with Allen wrench setscrew to loosen collar and adjust collar position

Some hoistway door unlocking device keys have a permanent stop that is either formed as a small tee on the key shaft or the design of the key physically limits how far the hoistway door unlocking device key can be inserted in the keyhole.

The pictures used to demonstrate operation of different types of hoistway door unlocking device keys are shown with ungloved hands. This is done for purposes of clarity. Ordinarily, gloves would be worn.

Fig. 13–15. Key is turned to unlock the door.

Drop Key (Single Link)

To position the single-link drop key it is necessary to align the key shaft and link with the round escutcheon collar mounted on the hoistway door. Slide the key through the keyhole until the link is free to drop on the hoistway side of the door. Rotate the key away from the leading edge of the hoistway door to engage the key unlocking bolt, paddle, or hoistway door drive roller(s). Just a small amount of force is required to unlock the hoistway door. If there is no movement when the link contacts the key unlocking bolt, then turn the key in the opposite direction. Figures 13–13 through 13–21 show drop key sequence of operation as viewed from landing and hoistway sides.

Fig. 13–16. Door is unlocked and moved to clear the interlock

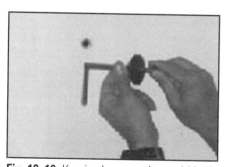

Fig. 13–13. Key is shown as it would look when in position on the other side of the hoistway door

Fig. 13–17. Key is removed and door is moved slowly to the fully open position

Fig. 13–14. Key is aligned with the keyhole.

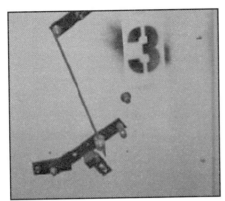

Fig. 13–18. View of escutcheon with tip of key

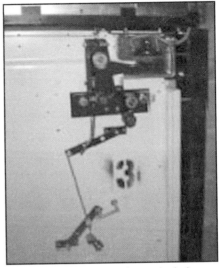

Fig. 13–20. Key is turned to unlock door

Fig. 13–19. Key in position to engage door-unlocking linkage

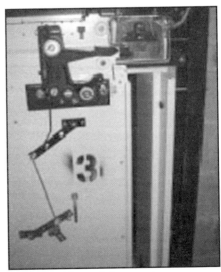

Fig. 13–21. View of door partly open after clearing interlock

If the drop key is inserted through the escutcheon collar and is rotated but does not meet with any resistance, then reposition the link closer to the hoistway door. This should allow contact with the key unlocking bolt. As a rule of thumb, when the link of the drop key is heard striking the face of the hoistway door as it falls into position, it is probably at the right distance from the door to engage the key unlocking bolt.

In the situation where the keyhole does not have an escutcheon and the drop key is inserted and rotated without engaging the key unlocking bolt, it is possible that a lunar key is required to unlock the hoistway door by using a downward motion. As emphasized earlier, little force is required when using a hoistway door unlocking device key such as a drop, lunar, or tee, where the key applies a force through rotation, downward motion, or side-to-side movement.

Drop Key (Center Link)

A variation of the drop key is one where the link is hinged near the middle of the key shaft. This special design is for the situation where the elevator car is at or near the landing of the hoistway door that requires opening. When this occurs, the typical drop key with a single or even a multiple-drop link will not work to open the hoistway door. This is because the elevator car's clutch has engaged the hoistway door drive roller(s), preventing needed movement to unlock the door.

Using a drop key with a center-hinged link usually solves this problem. The key is aligned with and inserted through the keyhole contacting the clutch of the car door. Forward force is applied to the key, causing the straight end of the drop key to depress the clutch, releasing the hoistway door drive roller(s). While holding the clutch in this position. the drop key is rotated to allow the link to contact and move the roller(s) to unlock the hoistway door.

Drop Key (Double or Triple Link)

Some drop keys have relatively short double or triple links that are hinged together. This type of key is useful when attempting to open a hoistway door where the elevator car is at or near that landing—but not in

range of the elevator car door's clutch. In this situation, the clearance between the hoistway and car doors is often too narrow to allow the single link of most drop keys to clear the opening and drop freely to its operating position (fig. 13–22).

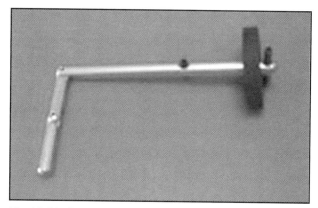

Fig. 13–22. Double-link drop key

The drop key is aligned with the keyhole, then inserted slowly to allow each link to clear the narrow space between the hoistway and car doors. After all links have dropped into position, the key is rotated to engage the key unlocking bolt. The action of the hoistway door unlocking device key lifts the key unlocking bolt and lift rod together to unlock the hoistway door. A limitation of the multiple links versus the single link is that use of the multiple links is not as rigid as a single link. As a result, when the multiple links contact the key unlocking bolt or paddle, the links tend to slide off the paddle.

Lunar Key

Another common type of hoistway door unlocking device key is the lunar key (also know as the half-moon key). This type of key gets its name from the U shape of one end of the key (fig. 13–23). There is a large and also a small lunar key. Usually the hoistway doors equipped with a keyhole for use by a lunar key has an escutcheon plate with a U-shaped cut in it. However, for a small lunar key, the keyhole is cut into the door as a small lunar or crescent shape. It typically does not have an escutcheon plate. However, one may be required if the center portion of the U has been broken off, as that would be a code violation.

Fig. 13–23. Large lunar (half-moon) key

To use the large lunar key, the shaft of the key is aligned with the lunar escutcheon plate and inserted about three or four inches (figs. 13–24 and 13–25). This should provide enough room to position the key under the key unlocking bolt. The key is pulled down using little force only. Once the lock is in the open position, slowly slide the door just enough to clear the interlock. Then remove the key and continue slowly opening the door to its fully open position (figs. 13–26 and 13–27).

Fig. 13–25. Key inserted about 3–4 inches

Fig. 13–26. Hoistway view of key

Fig. 13–27. Hoistway view of unlocked door

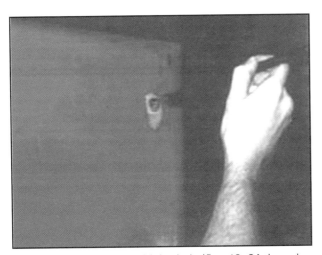

Fig. 13–24. Key aligned with keyhole (figs. 13–24 through 13–27 courtesy of John Fisher)

The small lunar key is either made of steel or an aluminum alloy (fig. 13–28). In the latter case, the key is subject to bending when too much force is used during its use. This key is operated in the same manner as for the large lunar key. The small lunar also can slide through the large lunar escutcheon plate to unlock the door.

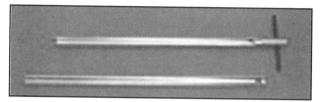

Fig. 13–28. Two types of small lunar keys

Lunar Key Operation (Variation)

A variation in the use of a lunar key is found where the escutcheon plate is mounted on an adjacent wall between or next to the hoistway doors (fig. 13–29). Unlike the pull-down motion by the large lunar key described earlier, the large lunar key is used to apply a force that is perpendicular to the face of the escutcheon plate (fig. 13–30).

Fig. 13–29. Lunar keyholes installed on wall between two elevators

Fig. 13–30. Lunar key pushed into wall to unlock door

While tension is maintained on the key to keep the lock in the open position, another firefighter is usually required to slide open the hoistway door. If a lunar key is not available, consider using a short length of steel electrician's tape. If the tape is not available, consider removing the escutcheon plate and using a thin-bladed screwdriver.

Tee Key

This type of hoistway door unlocking device key gets its name from the shape of the working end of the key. It looks like an inverted tee (fig. 13–31).

Fig. 13–31. Tee key

The escutcheon plate is also shaped as an inverted key. Like other keys, the escutcheon plate helps to facilitate positioning and use of the tee key.

Depending on the type of interlock or hoistway door release mechanism used, the hoistway door is unlocked by the application of forward or perpendicular force to the hoistway door provided by the tee key or by a downward or sideward motion of the key. Figures 13–32 and 13–33 show the tee escutcheon and use of tee key to unlock the door using a downward motion.

Figures 13–34 through 13–37 show operation of the key with a sideward movement. The tee key is used to unlock another type of door release mechanism, as viewed from the hoistway (figs. 13–38 and 13–39).

Fig. **13–32.** Tee escutcheon

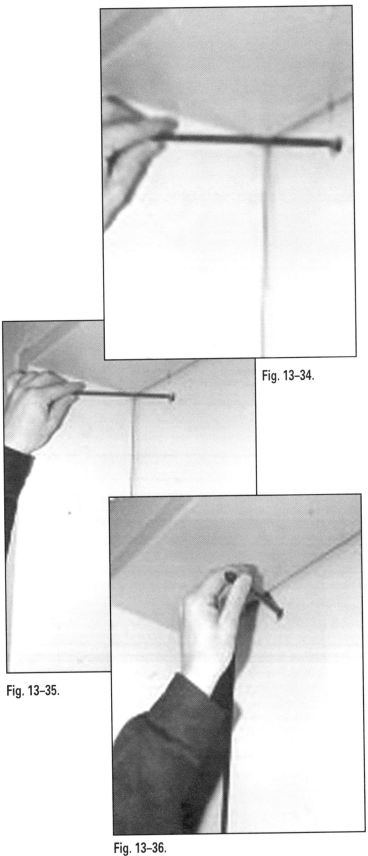

Fig. **13–34.**

Fig. **13–35.**

Fig. **13–33.** Tee key used to unlock hoistway door

Fig. **13–36.**

Fig. 13–37.

Fig. 13–38. Hoistway view of tee key in position to unlock door

Fig. 13–39. Tee key is pulled down to unlock door

Not all tee keys are of the same size. Although they are shaped the same, they might be too large to clear the escutcheon plate. If this is discovered during an actual incident and a lunar key is not available, consider using a short length of steel electrician's tape.

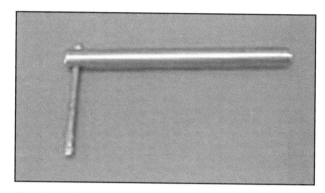

Fig. 13–40. Swing door key

The swing hoistway door has a keyway that is installed either on the doorjamb or the door itself near the leading edge of the door. The keyway is slotted to allow the tip of the swing key to engage the keyway. When the key is rotated a short distance, the door release roller arm, mounted on the hoistway wall directly opposite the keyway, is moved inward, unlocking the swing door (fig. 13–40). While the key is held in this position, open the door enough to clear the lock, and then open the swing door fully. Firefighters must exercise caution when using this key due to the closeness of the keyway to the leading edge of the swing door. Figures 13–41 through 13–44 showing swing door key ready for use, swing door keyhole and door lock, and hoistway views of door unlocking roller arm.

Fig. 13–41. Swing door key ready to use (figs 13–41 through 13–45 courtesy of Jim Jarboe)

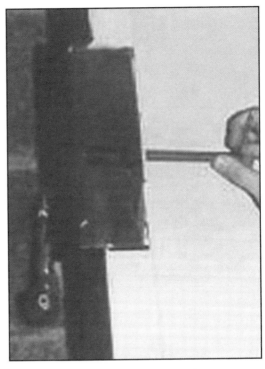

Fig. 13–43. Side view of key with roller arm open in locked position

Fig. 13–42. Keyhole located above door latch

Fig. 13–44. Side view of key with roller arm in unlocked position

Hook (Extrusion) Key

Some hoistway doors have a keyway mounted on an adjacent wall. The keyway has a recessed metal disc with a slot (fig. 13–45). A hacksaw blade is used as a makeshift hook key (fig. 13–46).

Fig. 13–47. Hook key next to slotted disc (figs. 13–47 through 13– 50 courtesy of Jim Jarboe)

Fig. 13–45. Slotted keyway located on wall next to elevator door

Fig. 13–46. Makeshift hook key made from a hacksaw blade

Fig. 13–48. Hook key engaging disc

The disc is under spring tension. Therefore, when the end of the hook (extrusion) key engages the disc slot, force is applied straight out from the wall to release the lock as shown in figures 13–47, 13–48, and 13–49. While holding the key in this position, the hoistway door is slid open just enough to clear the lock (fig. 13–50).

Fig. 13–49. The disc is pulled outward to unlock door.

After releasing and removing the hook key, the door is moved slowly to the full open position. Due to the distance between the face of the door and the keyway, a second person is usually required to open the hoistway door.

If the hook key is not available, do not attempt to use a screwdriver and fingers to pry open and hold the disc. Careless actions could result in painful injury should the disc slip and return to the closed position.

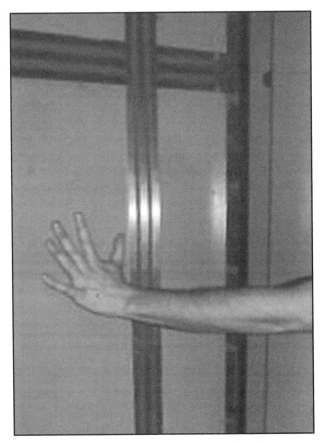

Fig. 13–50. While tension is maintained on the disc, the door is moved away from lock.

Freight Elevator Door Release Chain

On a freight elevator with vertical biparting doors, the hoistway door unlocking chain is protected and hidden by a locked box, mounted on an adjacent wall. Figures 13–51 and 13–52 are two types of devices. Figure 13–53 shows the device in the unlocked position.

Fig. 13–52. Keyed unlocking device box

Fig. 13–51. Keyed unlocking device box

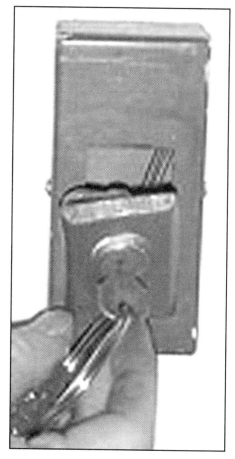

Fig. 13–53. Unlocking device box in unlocked position

A conventional-type key is required to unlock the box and allow access to one end of the chain that passes through a drilled hole in the wall where the other end is attached to the interlock release roller arm (fig.13–54). By pulling on the chain, the interlock roller arm is moved inward toward the hoistway wall, thereby unlocking the door (fig. 13–55). While holding the interlock roller arm in this position, open the biparting doors by slowly applying a downward force on the lower half of the doors. If available, another firefighter can open the door as described.

Fig. 13–55. Hoistway view of release roller arm. The arm is attached to other end of chain.

Firefighters should work with the building engineer to ensure accessibility of the key to unlock the door-release box. Without the key, firefighters should consider using a pike pole, elevator pole, or other suitable tool to manually unlock the biparting doors through movement of the door release roller arm. Remember, this is a nonforcing technique. However, if an emergency exists, then firefighters should consider forcing open the box to reach the pull chain.

Hoistway Door Access Switch

An access switch is a special key-operated switch *designed for sole use by an elevator mechanic or inspector.* It is not for use by firefighters. The purpose of this switch is to allow the elevator mechanic or inspector to gain access to the top and bottom of an elevator car for the purpose of inspection, maintenance, or repair. The key is not kept on the premises but rather carried by the elevator mechanic or inspector. The access switch is usually mounted adjacent to an elevator entrance at the top and lowest landings. The access switch is of no value in an emergency, because firefighters must be able to gain access to the elevator car to activate the access switch at the landing before using the key. It's both a matter of safety and potential liability.

Fig. 13–54. Chain is pulled straight out to unlock door. Biparting door is opened while tension is maintained on chain.

Improvised Hoistway Door Release Tools

In the early years (and even today), some firefighters used makeshift tools in an attempt to reach the hoistway door interlock and open the hoistway door. Yet, the use of these improvised tools can carry the risk of possible electrocution. This is reason enough for firefighters not to use an improvised tool as a substitute for a hoistway door unlocking device key. Moreover, the newer design of interlocks prevents the use of these potentially deadly tools. If a hoistway door unlocking device key is not available to open quickly a hoistway door, and an emergency exists, use of a forcible entry tool is a reasonable alternative.

List of Safety and Practical Points

To help reinforce the safety and practical points discussed earlier in the chapter, a list of items is provided here to serve to remind firefighters of the importance of following all safety principles and applicable practices not only when using hoistway door unlocking device keys but also during all elevator incidents. Practical hints serve to promote proficient and confident use of hoistway door unlocking device keys.

- Recognize the resourcefulness of elevator mechanics and inspectors. They are available to provide firefighters with useful information about different types of hoistway door unlocking device keys and their proper and safe use. Although some jurisdictions do not require that an elevator mechanic be on hand when firefighters want to open up a hoistway, it is a good practice to have one present.

- Regardless of whether it is a practical training session or a real elevator incident, be sure to remove power from elevators and use lockout/tagout safety procedures. The sudden and unexpected movement of an elevator car during a training session, or a routine or emergency elevator incident, can result in serious injury or even death. As an added sense of safety and where the top of the elevator car is at or within arm's reach, apply the emergency stop switch and place the

top-of-car operator in the inspection mode. Modern elevators will not allow this. After completing the training and before the main power is restored to the elevator car, return both switches to their normal positions.

- Hoistway door unlocking device keys should be part of the firefighters' elevator rescue tools and equipment. This ensures that the keys are available when needed.

- When using a building to train, be sure to notify and obtain cooperation from the building engineer or manager before using the hoistway door unlocking keys. They should give their approval because they are legally responsible for the building. After completing the training activity, notify the building engineer or manager.

- Assign one member of the training group to be the safety officer. He or she can monitor activities to ensure that the applicable elements of the four safety rules discussed in chapter 11 are addressed.

- Wear appropriate personal-protective clothing and use safety equipment such as head and eye protection, fall-protection harnesses and ropes, gloves, and safety shoes or boots based on training environment.

- Be careful not to unlock and open hoistway doors when children are present. After the training is completed and the firefighters have left, children may try to mimic how the firefighters opened the hoistway door by using a similar key or an improvised one. Injury can result from movement of the hoistway door or exposure to the deadly hoistway.

- During training, be sure to use the hoistway door unlocking device key at the lowest landing. If this is not possible, go to the next higher landing, then position the elevator car immediately below it. This is the safest way given that some pits are quite deep. If the fall hazard with the door open is more than 6 feet, fall protection should be worn per OSHA regulations.

- When operating the hoistway door unlocking device key, it is okay to slowly slide the door to the fully open position, provided the top of the elevator car that is located immediately below is at or very close to the landing used for training. A way to avoid injury is to remove the key from the hoistway door as soon as the key mechanically unlocks the door and the door keeper hook clears the interlock box.

- Use a door wedge tool to temporarily hold open a hoistway door. Avoid using a screwdriver, wedge-shaped doorstop, or clothespin. Also assign a firefighter to guard the open hoistway, and use a ladder to barricade it, if necessary. If a hoistway door is found with a broken door latch, stand guard to protect people, especially children, from getting too close to the door. Notify the building owner—it is their equipment. There may not be an elevator maintenance company to notify.

- When the hoistway door unlocking device key contacts the key unlocking bolt that is attached to the lift rod, apply just enough force to open the door lock only. Too much force can cause damage to elevator equipment or the hoistway door unlocking device key itself.

- Key damage is especially likely when the hoistway door unlocking key is made of a relative soft metal such as aluminum alloy and the key is used to apply force to the lock release through a vertical (downward) or horizontal (side-to-side) motion.

- While using the hoistway door unlocking device key to hold open the door release latch, slowly open the door just enough to clear the keeper. Then remove the key from the door and slowly open the door to its full open position. If the firefighter should forget to remove the key before opening the door fully, he will likely get a painful reminder when the hand gripping the key becomes wedged between the key and doorjamb!

- If too much force and speed are used when sliding open a hoistway door, it is possible that the door could jump its track and fall dangerously down the hoistway. This can be especially harrowing if the elevator car is positioned below.

- If some drop keys carried by firefighters have an adjustable stop collar, loosen the setscrew and move the collar closer to the tee handle end, then tighten the setscrew. This allows for more flexibility in using the drop key because the key unlocking bolt or other type of door release mechanism might be at different distances from the face of the hoistway door on the hoistway side. Therefore, with the stop collar relocated to the handle end of the drop key, the shaft of the drop key would be free to move the drop link closer or farther from the hoistway door to effectively engage the key unlocking bolt.

- When a hoistway door is returned to its fully closed position, be sure to check that the door locks properly. This usually requires pushing on the door, using a sideward motion in the direction the door opens.

- If the escutcheon plate is missing from the keyhole, check an adjacent or other nearby elevator hoistway door to identify the type of escutcheon plate that is used. If the escutcheon plate is missing and there is no nearby elevator to determine what type is used, first use a drop key and then a lunar key before trying other types of hoistway door unlocking device keys.

Usually the building engineer knows the location of the hoistway door unlocking device keys used in their building or group of buildings. The ASME A17.1 Code requires that a hoistway door unlocking key be provided in the building that is not accessible to the public but is accessible to emergency personnel.

- When using the swing door key, it is necessary to rotate the key about an eighth turn to engage and unlock the door release mechanism.

- When checking for the presence of a hoistway door keyhole, and if one is not found on any hoistway doors, check an adjacent wall of the elevator entrance. Sometimes the keyhole is installed on the lower part of a wall that is either located between two elevators or on an adjacent wall.

- Not all tee keys have the same width. Therefore, they may not fit all tee-shaped escutcheons. To help eliminate a potential problem, firefighters should carry tee keys of different widths.

- When conducting a training exercise in the proper use of hoistway door unlocking device keys, be sure that all firefighters use the keys. While this training is especially important to inexperienced firefighters, it also serves as a good refresher for members who have had earlier training in the use of these keys. It's all about teamwork and safety.

- If feasible, conduct a survey of your elevator buildings to gather information about hoistway door unlocking device keys, location of keyholes, and other useful information. Consider storing the information on a mobile data terminal (MDT) or in a notebook. Having this information readily accessible can improve operational efficiency.

Summary

Hoistway door unlocking device keys are tools available to firefighters to assist them in safely opening hoistway doors. The common types of hoistway door unlocking device keys are the drop, lunar, and tee. They are designed to unlock hoistway doors by key rotation, swing, push or pull. In most cases, little force is enough to unlock a hoistway door. Applying too much force to the key unlocking bolt or hoistway door release rollers not only can damage elevator equipment but also damage the key itself.

Hoistway door unlocking device keys might not be usable on some hoistway doors. The elevator safety code of some jurisdictions prohibits the use of hoistway door unlocking device keys primarily on the basis of the threat of envisioned public harm.

Through in-service inspection and periodic training, firefighters can gain useful knowledge to safely and effectively use hoistway door unlocking device keys. The local elevator mechanic is a great resource to firefighters. Include them in your training sessions and request them to respond to serious or complex elevator rescue incidents. The elevator mechanic can usually safely move an elevator to allow trapped passengers to walk out through the door at the landing. Use them whenever you can—it is safer for the passengers and the firefighters.

Good training promotes proficiency and instills confidence in firefighters. As a result, elevator hoistway door unlocking device key use provides for a safe and often effective way to open hoistway doors to reach and safely remove passengers.

The expression hoistway door unlocking device key is the terminology used in the elevator industry. However, it is common practice in the fire service to shorten expressions to facilitate communications during training and incident management. Again, if we teach them to use the correct terminology from the start, we don't have to worry about different terms meaning different things.

It is essential to reemphasize the importance of *power down* any time an elevator is used in training, or when managing an elevator incident. Power down is killing the power to an elevator by placing the mainline disconnect switch located in the machine room in the off position *and* following lockout/tagout procedures. Power down not only prevents an elevator car from moving, but also helps to prevent firefighters from getting injured or killed!

Review Questions

1. List the three most commonly used hoistway door unlocking device keys.

2. After using a hoistway door unlocking key to move the keeper hook away from the keeper, and the hoistway door has been moved clear of the interlock, what is the next thing you should do to avoid injury?

3. After using a hoistway door unlocking device key to unlock a hoistway door, and the door has been returned to its fully closed position, what should you do as an added safety measure?

4. What is a door wedge tool?

5. To prevent movement of an elevator car during training in the use of hoistway door unlocking keys, what should you do?

6. What are the two functions of an interlock?

Field Exercise

Schedule a training session with an elevator mechanic to practice how to use hoistway door unlocking device keys. Be sure to get permission to use the building from the owner or property manager before doing the training.

Chapter 14
Poling Guidelines

Introduction

Poling is a procedure that uses a long round or flat pole to reach and unlock a hoistway door from a landing (floor) above or next to it. The chapter discusses guidelines for using this procedure as well as the safety concerns. Firefighters are cautioned to wait for the elevator mechanic before using this procedure, unless circumstances and existing conditions indicate otherwise.

Poling across, down, and up are three types of poling operations. However, as described later, because poling up is not as safe as poling across and down, we do not recommend its use.

Safety Concerns

Working near or at an open hoistway is a dangerous situation. It is one that requires continued awareness and attention to safety by firefighters.

Poling requires that firefighters who are involved in rescue operations do the following:

- Wear head, eye, hand, and foot protection, and use fall-protection harnesses and ropes.

- Ensure that power is removed from elevator(s).

- Ensure that lockout/tagout procedures are done.

- Be sure a firefighter is assigned to guard the landing before and during opening of the hoistway door by poling. An opened hoistway door can expose the dangerous hoistway.

Moreover, ongoing communication and coordination is necessary among firefighters directly involved in the poling operations. The team leader needs to confirm and be notified when the elevator mechanic is en route.

Assigning a safety officer(s) and discussing the potential hazards attendant to working at or near an open hoistway is part of the safety plan. When conducting poling-down operations, assign a safety officer to each floor. The fall-hazard potential attendant to an open hoistway commands respect and due caution. Use a ladder to barricade the entrance to an open hoistway if the doors must remain open for more than a few minutes.

Begin poling operations on the floor nearest to where the stalled elevator is located. If poling down requires repeating the operation for several floors, then the incident commander must weigh the added risk against the desired outcome. A stalled elevator with entrapped passengers alone does not warrant the risk.

The poling tool is used to unlock a hoistway door, not open it. Adhering to this practice will help eliminate premature opening of a hoistway door before firefighters are in position to guard the opening.

Hands-on and supervised training using these poling guidelines will help firefighters become more proficient and confident. Only firefighters properly trained in the use of poling should use poles during actual elevator incidents. The rule here is simple: If you are not trained to use poling tools, don't use them!

Poling Tools

Pike pole

Aside from its routine use by firefighters to open walls and ceilings, the pike pole is used also to unlock a hoistway door. However, due to its inflexibility and relatively large diameter, the pike pole has limited use in poling across operations. In some situations firefighters cannot carry a pike pole to an upper or lower floor because the pole is too long to clear the stair or landing. In other situations, there is insufficient space at an elevator landing or between an elevator car and a hoistway wall to allow proper placement of the pike pole.

Another concern is the pike pole's weight. If a pike pole is used to unlock a hoistway door from a landing above, the firefighter controlling the pike pole might have difficulty reaching and unlocking the hoistway door. The degree of difficulty is also influenced by the pike pole's length and diameter. The longer the pole, the more challenging it is to control and position it in the hoistway.

Elevator pole

Some firefighters use a flat pole made of wood with a specially designed metal head called an elevator or flat pole (fig. 14–1). This pole is often used instead of a pike pole because it is more flexible and easier to use. The flat pole often is used to unlock a hoistway door from the

entrance of an elevator car located next to the hoistway door that requires opening. It is also useful when using poling down guidelines.

Fig. 14–1. A common tool used to unlock a hoistway door

In the early 1970s, while working for the Silver Spring (Maryland) Volunteer Fire Department, Ted Jarboe and Dean Groseclose codesigned an elevator pole. The pole was made of thin, flat wood and had a metal head designed to exert a pushing or pulling motion to unlock a hoistway door. The pole was and is still referred to as the JarClose elevator tool (pole) after its two designers. The elevator pole is used not only in Montgomery County, MD, where the pole was designed, but also in neighboring jurisdictions and in other states. It's proven to be a useful tool in unlocking elevator hoistway doors.

Door unlocking force

As when using a hoistway door unlocking device key, poling usually requires little force to unlock a hoistway door. Although it is relatively easy to apply the required force when poling across, it is not so easy when pulling down. In the latter procedure, the force applied to the hoistway door roller is at right angles to the length of the pole. The main purpose of the poling tool is to unlock the door, not physically slide the door to its open position. Firefighters stationed at that hoistway door's landing are responsible for slowly and safely opening the hoistway door and to guard the exposed hoistway.

Why Poling Doesn't Always Work

Although poling is an effective procedure, it doesn't always work. Obstructions in the hoistway, such as a column, wire mesh, or car door restrictor, can prevent effective positioning and use of the poling tool. In some situations, the opening between the front of an elevator car and a hoistway wall is too narrow to slide the poling tool into position, or the elevator car's clutch prevents the hoistway door release roller from unlocking. Still another reason is that sometimes the combined reach of the poling tool and firefighter is not enough to span the opening.

Poling Guidelines

Regardless of which floors they are working on, firefighters must exercise extreme caution. Power down, lockout/tagout guidelines, use of fall protection equipment, and guarding hoistway openings are basic, yet essential, safety measures. Communication and coordination is especially necessary among firefighters assigned to work together, though on different floors. Firefighters must never ignore the deadly threat of an open hoistway.

Poling up

Poling up was a method used in elevator rescue training. However, it was rarely used during an actual elevator incident and was later abandoned. The purpose of poling up was to open a hoistway door from a landing directly below it. Although firefighters did adhere to strict safety guidelines including, for example, power down (deenergizing) of elevators and use of fall protection, there was greater risk of harm compared to the risk associated with using other poling guidelines. There are safer options available to firefighters such as using a hoistway door key, or simply waiting for the local elevator service company representative. For these reasons, we do not recommend that firefighters use the poling up procedure.

Poling across

Poling across is a procedure that uses a poling tool in a horizontal position to reach and unlock a hoistway door from a landing that is next to the hoistway door that requires opening. Poling across is commonly used in high-rise buildings and other elevator buildings that have one or more multiple-car hoistways. Figures 14–2 through 14–5 show the pole being used to unlock the hoistway door by applying a pushing motion.

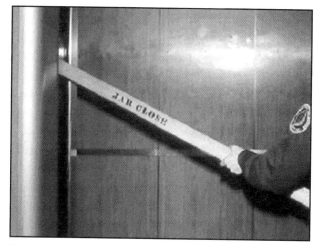

Fig. 14–2. Elevator pole being positioned from the rescue-assist elevator car

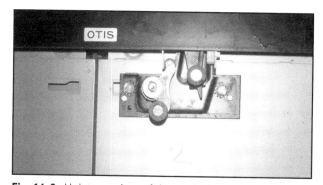

Fig. 14–3. Hoistway view of door unlocking roller

Fig. 14–4. Pole tip in contact with door release roller

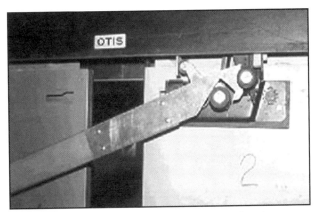

Fig. 14–5. Pole is pushed to unlock door and door is slightly opened.

In figures 14–6 through 14–8, the pole uses a "pulling" motion to unlock the hoistway door from the opposite direction. Figure 14–9 shows a variation of the door-release mechanism. In this case and where there is not a keyhole, the pole can open the door by moving the paddle.

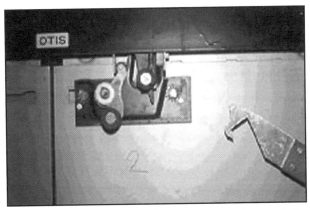

Fig. 14–6. Pole is positioned from other car to demonstrate pulling motion.

Fig. 14–7. Pole hooks the unlocking roller.

Fig. 14–8. Pole is pulled to unlock and open hoistway door.

Fig. 14–9. A variation of the door-release mechanism

The poling tool is used to apply force to the hoistway door roller in the same direction as movement of the tool. After the landing is identified where poling operations will start, the elevator car serving that landing is brought to the floor and the main power to it removed. Applying the emergency stop switch located inside the elevator car if available is another safety precaution.

A firefighter team is stationed in front of the hoistway door that requires opening; another firefighter gets ready to use the poling tool from the adjacent elevator car entrance. While standing at the entrance of the elevator car, the firefighter looks across the opening to locate the hoistway door drive roller of the adjacent hoistway door. If there is insufficient lighting to locate the hoistway door roller, assign a second firefighter to direct the flashlight.

After receiving the poling tool and alerting the team guarding the hoistway door that is about to be unlocked, the firefighter extends the pole through the opening to reach and unlock the door. As the firefighter holds the door in the unlocked position, they tell the firefighter team to slowly open the door. The firefighters at the opening hoistway continue to guard the opening. Then the other firefighter removes the poling tool, releases the stop switch, and closes that hoistway door.

If other hoistway doors require opening, then consider the guidelines described below for poling down. The power should remain off to both elevators. After completing the rescue operation, do not restore power to the stalled elevator. Depending on circumstances, power may be restored to the other elevator. When in doubt, leave the power off.

Poling down

When a poling tool is used to unlock a hoistway door from a landing above, the procedure is commonly referred to as poling down. This method is more challenging than poling across. It requires that firefighters work from an open hoistway, position the poling tool in the hoistway, and apply a force at right angles to the poling tool to unlock the hoistway door. A poling down sequence is shown in figures 14–10 through 14–13. Figures 14–14 and 14–15 show a different arrangement of the unlocking mechanism (roller). Firefighters may encounter different unlocking mechanisms; however, in principle, they operate the same. When using the pole, firefighters can quickly determine how to move an unlocking mechanism to release the hoistway door by using a trial-and-error method. Typically, when two rollers are found one above the other and offset, the top roller is the unlocking one.

Fig. 14–10. Firefighter positions pole in hoistway (photo by Robert Wilcox).

Fig. 14–11. Pole engages door release mechanism (photo by Robert Wilcox).

131

Fig. 14–12. Pole unlocks hoistway door and door slightly opened (photo by Robert Wilcox).

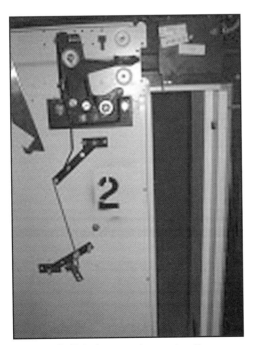

Fig. 14–13. Pole is moved away to allow firefighter in hallway to fully open door (photo by Robert Wilcox).

Fig. 14–14. Pole is positioned to unlock hoistway door using different type of door release roller. (Note: Keeper hook is in closed position.)

Fig. 14–15. Pole lifts release roller to unlock door. (Note: Keeper hook is in open position.)

Fig. 14–16. With this type of unlocking mechanism, the pole moves the paddle to unlock the hoistway door.

Firefighters may be able to gain access to the hoistway designated for use in poling down by either using a hoistway door key, if a keyhole is provided, or using the poling across procedure from an adjacent elevator car. For safety reasons, poling should begin at the landing nearest to the stalled elevator car.

After addressing safety issues, stationing firefighters at both landings, and assigning another firefighter to hold open the hoistway door, the two firefighters of the poling team lie face down on the landing floor and position themselves near the hoistway with only their heads and arms exposed to the open hoistway. One firefighter is given the pole, and moves it down the hoistway and waits, while the other firefighter uses a flashlight to locate the door release roller. Once the roller is located and the team on the lower floor is alerted, the pole is moved sideways to contact and unlock the door. The force applied to the roller is at an angle to the length of the pole. This maneuver can be difficult to accomplish, depending on the pole length and position or configuration of the roller.

After getting the okay from the poling person to open the door, firefighters on the floor below slowly open the hoistway door and guard the opening. If the desired landing is reached, then the poling team can remove the poling tool and close their hoistway door.

At this point, poling operations are complete, and equipment no longer needed is safely set aside for later return to the emergency vehicle. Firefighters continue rescue operations using other guidelines to remove passengers from the stalled elevator car.

Passing the pole

If firefighters need to open more than one hoistway door before reaching the desired floor, the firefighter on the poling floor usually can pass the pole directly to the team on the floor below. However, before the pole is passed to the firefighters on the floor below, they should take a prone position on the floor similar to the position taken by the poling team.

After alerting the team below, the firefighter carefully lowers the poling tool by hand. However, if the safety officer or the team receiving the poling tool thinks it is too dangerous for a direct handoff, then the poling tool is removed from the hoistway and carried downstairs to the other team. This pole-passing procedure is repeated

until the hoistway door nearest to the stalled elevator car is opened. As a hoistway door is opened and closed, it is checked to confirm that it is locked.

General Poling Points

These points illustrate a general sequence of poling operations. The explanation starts at the point in the rescue operations where poling is selected as a means of unlocking one or more hoistway doors. It ends with opening of the last hoistway door, removing the poling tool, and securing the elevator car and hoistway door on the above landing.

- Assemble team at floor where poling operations will begin.

- Ensure that the poling tool and other elevator rescue equipment are ready for use.

- Review potential hazards and safeguards.

- Review the poling plan.

- Assign teams to power down and lockout/tagout elevator(s); open, guard, and barricade hoistway doors; use the poling tool; and where needed, direct the flashlight.

- Assign a qualified firefighter to serve as the safety officer.

- Ensure that members have the appropriate personal protective equipment and portable radios, consistent with their respective assignments.

- Ensure continued communications and coordination among teams.

- Allow just enough firefighters to safely work at or near a hoistway opening.

The following list highlights or adds to safety and procedural considerations noted earlier. Though not necessarily complete, the list can assist firefighters who are developing or revising their poling guidelines. Firefighters also can use these points to develop a poling quick-review guide to carry on emergency vehicles.

- Recognize the resourcefulness of the elevator mechanics and inspectors. Many jurisdictions legally require that an elevator mechanic be on hand if we want to open up a hoistway.

- Regardless of whether it is a practical training session or a real elevator incident, be sure to remove power from elevators and use the lockout/tagout guidelines. The sudden and unexpected movement of an elevator car during training or an elevator incident can result in serious injury or even death.

- When training, be sure to notify and obtain approval from the building engineer or manager before beginning poling operations. He or she should give the okay since that person is legally responsible for the building. After completing the training activity, notify the building engineer or manager.

- Assign one member of the training group to be the safety officer. If poling down, assign a safety officer to each floor. The safety officer can watch for hazardous conditions and monitor firefighters to ensure continued use of safe practices.

- Wear head and eye protection, fall-protection harnesses and ropes, gloves, and safety shoes or boots.

- Before actually beginning poling operations, firefighters assigned to use the poling tool, flashlight, and guard the hoistway opening on the floor below should check the tightness of their hardhats or helmets to prevent loss down the hoistway. The sudden loss of this safety equipment can be a dangerous distraction to the wearer.

- Avoid wearing loose-fitting clothing when working close to elevator equipment with moving parts.

- Do poling across training to open a hoistway door that is located directly above its elevator car. The adjacent elevator car is relocated to the same floor as the target hoistway door that is to be poled open. Both elevators are powered down, including lockout/tagout procedures. The pole is placed between the front of the elevator car and hoistway wall to reach the hoistway door to unlock it. Poling across is a safer method than poling down operations on the floor immediately above an elevator's pit, because some pits are quite deep.

- Do poling-down training to unlock no more than two hoistway doors. Locate the elevator car at the landing that is immediately below the landing of the hoistway door(s). Two hoistway doors are recommended when firefighters want to practice the pole-passing procedures.

- Limit the number of firefighters working or standing near an open hoistway.

- Use a door wedge tool if it is necessary to hold open a hoistway door. Avoid using an item such as a screwdriver, wooden wedge, or clothespin to hold open a hoistway door. The door wedge tool was designed for use by elevator mechanics and inspectors. Although firefighters should consider adding a door wedge tool to their elevator rescue equipment inventory, its use does not eliminate the need to have a firefighter standing near the door for as an added safety measure.

- Use a ladder to barricade entrance to an open hoistway.

- Never leave a hoistway door unguarded.

- Notify the building manager or engineer if a hoistway door is found with a broken door latch. Firefighters should stand guard until the manager or engineer arrives and takes appropriate measures to guard the door until it is repaired.

- Use the poling tool to unlock the hoistway door only. Let the other assigned firefighter team move the door slowly to the open position. Adhering to the limited use of the poling tool will help to prevent premature opening of a hoistway door and exposing the unguarded hoistway.

- Check to ensure that the hoistway door is locked after it is closed.

- Be sure that all attendees participating in poling training use the poling tool(s). While this training is especially important to inexperienced firefighters, it also is a good refresher for members who have had earlier training in the use of poling tools. It's all about teamwork and safety.

Poling Open a Freight Elevator's Biparting Door

Basically, the principles applied to poling open passenger elevator hoistway doors also apply to poling open biparting hoistway doors. However, the freight elevator doors use a door release roller arm instead of door drive roller(s) as the unlocking device. Both devices are connected to their respective interlocks. The biparting door is unlocked when the door release roller arm is moved sufficiently inward toward the hoistway wall. This action moves the keeper hook away from the interlock keeper.

Some firefighters use a special tool designed to unlock biparting hoistway doors. This L-shaped tool is inserted through the space where the two halves of the biparting hoistway doors meet, then moved across into the hoistway to reach and pull inward the door release roller arm—to unlock the door. While one firefighter maintains force on the poling tool to keep the lock open, another firefighter manually opens the biparting door. Sometimes a strap is attached to one of the biparting panels to facilitate opening.

Firefighters can get instruction in how to safely unlock and open biparting hoistway doors from an elevator mechanic, qualified building engineer, or a qualified member of the fire service. Before practicing guidelines, firefighters must obtain approval from the building manager or engineer. Even when a firefighter qualified to instruct poling guidelines for biparting freight elevators is present, it is advisable to have the building engineer or an elevator mechanic present as well. It is not uncommon to have an elevator fail to operate during training sessions. When this happens, the elevator mechanic can correct the problem and restore the elevator to service.

Guidelines in Review

Opening doors from same level (poling across)
- Disconnect main power to elevator and use lockout/tagout procedures.
- Apply manual force to the upper section of the biparting hoistway doors to provide space to insert a poling tool, such as an L-shaped tool.

- Slide tool between both sections of the hoistway door and extend it into the hoistway near where the door release roller arm is locked.
- Angle poling tool or L-shaped tool to engage arm of door release roller.
- Apply force to move roller arm in towards hoistway wall.
- Use a second firefighter to separate biparting hoistway doors.
- Remove the poling tool and guard open hoistway for remainder of rescue operations.

Opening doors from landing above (poling down)
- Disconnect main power to elevator and use lockout/tagout procedures.
- Use key to the access chain attached to roller arm to unlock doors if available, or follow the guidelines above.
- Wear head and eye protection, fall-protection harnesses and ropes, gloves, and safety shoes or boots. Follow other safety guidelines specific to poling down, as described earlier in the chapter.
- Position the poling tool to contact door release roller arm.
- Apply force to move roller arm in toward hoistway wall.
- Holding the roller arm in this position, alert firefighters on floor below to open the biparting hoistway doors. After doors are opened, remove poling tool, close doors, and check to confirm that they are locked.
- Continue to guard exposed hoistway on floor below until operations from that landing are complete.

Summary

Poling is another procedure available to firefighters to unlock hoistway doors. The procedure uses a poling tool to unlock a hoistway door either from an elevator car entrance next to or from the landing immediately above the hoistway door. When the poling tool is used between two elevator car entrances, it is referred to as poling across. Poling down is the expression used to describe operations where the poling tool is used from the floor immediately above.

In either case, the poling tool is used to unlock a hoistway door, not move it to its full open position. The reason for this is to prevent a hoistway door from being opened prematurely and exposing the unguarded hoistway opening.

Usually, little force is needed to unlock a hoistway door by either poling method. However, poling down requires that force be applied at an angle to the pole. As the required pole length increases, the difficulty in applying the sideward force also increases. Communication and coordination between the poling team and the team that opens the hoistway doors are crucial.

As with any guidelines, firefighters engaged in poling operations must power down elevators, guard and barricade any exposed hoistways, and use fall protection and other protective equipment. An added precaution is that firefighters directly engaged in poling-down operations must lie face down with only their heads and arms extending into the hoistway. This position helps to minimize fall hazard and promote confidence among firefighters using the poling procedure.

Firefighters should not use poling guidelines if safer methods are available, such as use of a hoistway door key. The elevator mechanic is not only a valuable resource during training but also during actual rescue operations. However, if a life-threatening elevator emergency exists, and firefighters must act before arrival of an elevator mechanic, then poling is a reasonable option.

Review Questions

1. Define the term poling.

2. What is the main purpose of the poling tool?

3. List three reasons why poling may not work.

4. List three safety precautions that firefighters should take before using the poling-across procedure.

5. List three safety precautions that firefighters should take before using the poling-down procedure.

6. Before using a door wedge tool, what must you do to ensure that the opened hoistway is guarded?

Field Exercise

Schedule a training session at one of your elevator buildings or parking garages to practice poling-across and poling-down guidelines. Be sure to include the elevator mechanic as an instructor resource to provide guidance and supervision.

Chapter 15
Car Top and Side Emergency Exits

The most dangerous operations that we as firefighters may encounter are ones where we are forced to operate on the top of an elevator in the hoistway. The modern elevator may have a handrail on the side or back of the elevator to provide a mechanic with a measure of safety, but this is not to be found on many of the older units that are out in the field today. In the past, members of the fire service may have operated the car with the top-of-car controls, but this is no longer allowed (fig. 15–1). The design of this system now requires a complex set of operations to be followed by the mechanic that are not available to the firefighter. We must remember who we are, and that we are visiting a dangerous place that is not our normal operating environment and presents great danger to us and the public we seek to help.

Top Emergency Exit

Fig. 15–1. An Otis car top, car door operator, double-slung ropes with side guard rail, car top operating control box

Follow the rules of the chapter 11, Elevator Safety, and *never* move a car without a mechanic's direct supervision. This is a most critical decision that should be thoroughly evaluated by the incident commander (IC) and the safety officer (SO) before members are committed to entering the hoistway. The IC and SO must weigh the needs versus the obvious and not-so-obvious dangers of placing the firefighters as well as the public in this dangerous atmosphere. The question is, why *can't* we wait? All of the variables must be examined, as to what the needs of the incident actually are. Passengers are safe, "locked in the box," until we arrive. Now, we urge them and the uninitiated firefighter to walk the tight rope of a car top, leaving the safety and security of the box, exposing them to the clutter and darkness—and the possible death of a fall into the hoistway! We must ask these questions first:

- Can the situation allow us to wait?

- If not, what other options exist?

- Have all the safety concerns been examined and assigned as a task?

- Are there sufficient and proper help, lockout/tagout procedures, ropes and harnesses, EMS, and the like?

- What is the ETA of the elevator mechanic?

- Last, ask yourself the *first* question again!

Guidelines

In earlier sections of this book, we listed the process and guidelines that we should follow when embarking into this dangerous area. The rooftop of an elevator is not a very hospitable place, with crosshead beam, car door operator motor(s), car top inspection station, 2:1 sheave setups, and poor footing with tripping and fall potential. This is enough to make the job of the safety officer (SO) at these types of calls the most critical staff assignment that the incident commander (IC) will make.

When first examining a car top of a stalled elevator, pay strict attention to any evidence of slack rope accumulation on the car top or in the machine room.

This will be an indication of an extremely dangerous condition, possibly caused by the car being on the safeties. We all have read of incidents that have gone into disaster mode, and could see where none of these questions was asked by the people in charge, or their fellow firefighters. After the fire is over, we all ask the right questions, except they do not do anything to help the situation at that point.

Fig. 15–2a. Car top exit, safety circuit plug, key lock (seismic rule)

Fig. 15–2b. Angled car top hatch

Car Top and Side Emergency Exits

The exit shown in figure 15–2a is shaped like a triangle, and usually located at the rear aspect of the top of the car. It may open in the area of the refuge space that is mandated as part of the safety considerations provided for the mechanic. It might be the center or left or right quadrant, depending on the design of the system. ASME A17.1 requires that the top emergency exit opening have an area of not less than 400 square inches and measure not less than 16 inches on any side. Usually a top-of-car emergency exit is of a square or rectangular appearance, as shown in figure 15–3 The code normally requires them to be locked from the outside to prevent occupants from getting out and on top of the car and becoming a statistic in *Elevator World* magazine. This is done by a number of features. Depending on installation date, and the edition of the code that it was installed under, the hatch will be protected by one of the following:

- Seismic Risk Zone 2 or greater (Rule 8.4.4.1.1) key tumbler and a car door electric contact (Rule 8.4.4.1.2), the opening of which will limit the car speed to not more than 0.75 meters per second or 150 feet per minute (ft/min). This rule is intended to allow a mechanic to view the hoistway and counterweight rails, looking for damage that could cause an impact between a moving elevator and other equipment after a seismic event.

- A variety of wing nuts, window sash latches, or slide bolts latches, which must be operated from the outside, from the roof of the car.

- At least one authority having jurisdiction (AHJ) has made its common fire service key also its seismic key.

The picture in figure 15–3 shows a top hatch opened, with a light baffle still in place. This is easily removed and passed either down into the car or out onto the floor landing. Others may be secured to the car top by a chain and must be worked around. It is obvious that it is easier to work with a light rather than in the dark of the hoistway. The photo in figure 15–4 shows a hatch opened with a folding or collapsible ladder extended down into the car to the occupants.

Fig. 15–3. False ceiling in car

Fig. 15–4. Escape hatch with ladder

Whenever we remove the occupants, all must be restrained with the proper rope and harness for fall protection, as seen in figure 15–5.

Fig. 15–5. Safe removal with ropes and harness

The protection of the public and the firefighters is an ongoing job that must be evaluated on a continual basis until the incident has been terminated. When finished, it is the IC's responsibility to make sure that all openings into the hoistway have been closed and rechecked for closure and security. If it is not possible due to the nature of the incident, then security must be provided by the fire service until responsibility can be accepted by other authorities who actually can perform that function. Be sure to leave the system under the watchful eyes of a qualified responsible party who will safeguard the area.

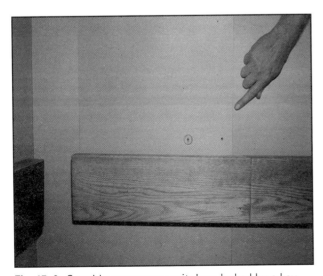

Fig. 15–6. Car side emergency exit door, locked by a key and tool

Car side exits are another issue altogether. Many of us, including both authors of this book, have used side exits to evacuate people from elevators. The common practice was to open the side exit with a small key that management would have, then press the contact for the interlock that was positioned on the side of the elevator (fig. 15–6). This interlock was mounted on the outside of the cab, exposed to the dirt and other trash that flows around in the hoistway due to the piston effect of the moving cars. The bottom line is that the interlocks were not reliable, and many people who depended on them to open the safety circuit were killed or seriously injured when the car, with its side door open, responded to a floor call. ASME-A17.1a-2002/CSAB44 update No. 1–02 sealed their fate when it prohibited the installation of side exits. In fact, they had not been a requirement since the publication of the ASME-A17.1–1991 edition of the code. To further ensure the nonuse of these doors, many AHJs ordered them to be permanently sealed. Their rationale is that it is safer to remove passengers through the normal door opening when possible, followed by the car top exit if the mechanic cannot move the car to allow the former.

The photo in figure 15–7 shows a young teenager who was caught by a moving car as a fire department tried to transfer him from one car to another. *They did not shut off the power at the main power disconnect!* When the car responded to a floor call due to a defective side door interlock, it pinned him between the car floor and superstructure steel out in the hoistway between the cars. Finally, after the firefighters kept pulling on him, he was pulled into the car. It then traveled 15 stories, reclosed its doors, and went 15 stories back down to the main lobby. This boy is lucky to be alive, in spite of the fire department's actions. They responded with the best intentions, had a department standard operating guide (SOG), but it would have been better if they had stayed in quarters. They certainly meant no harm, but came close to killing him. Their excellent SOG was not followed in any way.

Shown in figure 15–8 is an atrium style elevator, with a guardrail around the perimeter. There are no car top exits in atrium elevators, unless the hoistway is enclosed by glass entirely, due to the extreme danger the roof of these cars present to anyone on them, including elevator mechanics. In closing this chapter, it should be clear to all readers that the prescribed means of removing passengers from elevators is by the elevator car door.

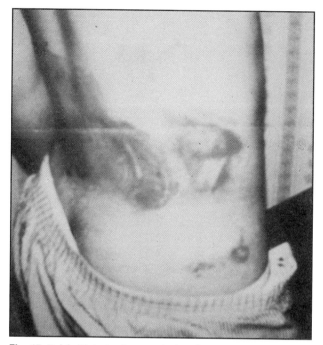

Fig. 15–7. Multiple bruises and abrasions on teenager caught by a moving car

This may require firefighters to wait for the mechanic, who can either move the car electrically or manually. The next opening would be the car top exit, but we have all moved into an area of danger to all concerned. *The use of side exits is prohibited.* In the past, they hanged pirates who forced the unfortunate to walk the plank over the abyss. We, as firefighters, go there to help them, not to endanger them. In the future edition of ASME *17.4–2005-Emergency Evacuation Guide*, the use of side exits is prohibited. The entire section has been removed from that standard.

Summary

This chapter is an extremely important one for the safety of responding firefighters. The top of an elevator car is one that firefighters do not visit very often. In the past, it was possible to operate a car from the control box located there. The elevator code does not allow this any longer, and it is *not* considered to be a safe practice for firefighters to perform. By following the rules of safety as seen in chapter 11, we can safeguard ourselves from injury. Firefighters never move an elevator, unless under the direct supervision of an elevator mechanic.

A critical decision must be made after an evaluation of the incident by the assigned IC and SO. The big question that must be asked and answered is, *why can't we wait?* Unfortunately, the time has not been taken in some incidents, with bad outcomes.

If the decision is that entry must be made, then other observations must be made. These include looking for slack rope before stepping foot onto the car and locating the emergency top exit.

The emergency top exit will be locked from the outside, but if a key tumbler is used, it is for Seismic Risk Zone 2 or greater. Last, this brings us to the side exit in the car. These will not be found in modern elevators, having been prohibited in 2002. The existing ones are to be permanently sealed from the outside to prevent their use.

Fig. 15–8. Atrium car

Review Questions

1. Are firefighters allowed to operate the car from the car top control box?

2. What question should be asked before entry into the hoistway?

3. What fire officers on scene should make that decision?

4. Why do we *not* use side exits?

Field Exercise

Establish links with a local elevator company and arrange for a tour of an elevator car top. During this session, locate the emergency top exit and the refuge space on the car.

Chapter 16

Forcible Entry: Types, Use, and Safety

During the preceding chapters, we have focused on the varied ways that we can find access to the passengers trapped in the elevator. Whether we use the hoistway door unlocking device key or rely on the assistance and knowledge of the elevator mechanic, there will be times when forcible entry must be used.

Fig. 16–1. Staffing and assorted equipment for forcible entry

Forcible entry is the last resort in the normal "locked in the box" elevator incident, but it may be our first choice in a situation that is a true emergency. Assembled in figure 16–1 is an array of forcible entry equipment, all of which have been used by fire departments in various types of responses. The decision to perform forcible entry can result from any of the following incidents:

- A medical emergency

- An entrapment outside of the car, but within the hoistway

- A fire situation requiring immediate removal

- A prolonged response (2+ hours) by a mechanic due to extensive travel or time delay

- A car with no response from within, suggesting a life-threatening situation

Whatever our situation is, we are going to follow a set of rules as we do with any elevator incident, namely, the four rules of safety that we learned in chapter 11, Elevator Safety. The first action that we will take is to shut off the power at the mainline disconnect in the machine room. As a precaution, since we are entering the hoistway, we will also bring the car(s) on either side of the stalled car down to the level we will be working at, and shut off the mainline disconnect to each. As we perform lockout/tagout on each disconnect, we should also maintain a fire department presence in that machine room. This is an additional level of safety for our members, to prevent "Ernie" from coming into the machine room and removing our lock and turning the power back on. A well-prepared and well-trained fire department will make sure that the proper forcible entry tools have been brought to the incident when entering the building.

Our situation is evolving in this manner:

- An elevator is stalled with an unresponsive occupant.

- The elevator mechanic is not on scene, with prolonged estimated time of arrival (ETA).

- Power to the affected elevator(s) has been locked out/tagged out at the mainline disconnect.

- The decision is made to use forcible entry.

The Irons

The most basic set of tools in the fire service are the "irons." Many elevator people refer to this set as the other fire service key. The irons includes two tools: the flat head axe, which is used for blunt impact force when used with the Halligan tool. As a set, they are used every day in all fire departments across this continent.

The Halligan, shown in figure 16–2, is a single tool that combines the features of the hook, axe, Kelly, punch & chisel, and claw tool. It was invented by Hugh A. Halligan in 1948, who served with the Fire Department of the City of New York (FDNY) from 1916 to 1959. This tool is universal, and can be seen in every Engine Company, Ladder Company, and Rescue Company that have the words *fire department* on its apparatus door.

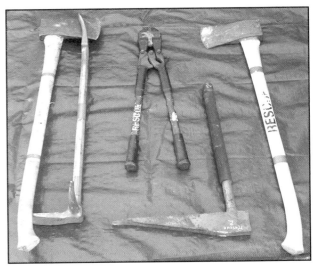

Fig. 16–2. The irons

Fig. 16–3. The irons in use

When the decision has been made to use forcible entry, a plan must be put into practice to assure a successful outcome. The placement of the Halligan tool is critical to making this incident one that results in a quick and safe removal, not one that will be talked about at the kitchen table for years to come. The location of the interlock must be determined before the bar is placed on the door. This can be done by:

- Seeing which way the door slides. If there is a hallway on one side of the elevator, then the car door will not have enough room to slide open in that direction.

- Slide a piece of card or other thin object into the opening between the surfaces of the door where it meets the wall frame.

- If the paper keeps going into the space, then that is the direction that the door opens into.

- If the paper stops, then you have hit the jamb where the closing floor door stops. The door opens in the other direction. Your Halligan goes at this end, as high up into the corner as is possible (fig. 16–3).

- *Never* place tools lower than a firefighter's shoulders. Any force applied lower than that, or at the bottom of the door, will be counterproductive and result in a "teepee" opening effect on the floor door. It will hang there, unusable, dangling into the hoistway, creating a dangerous situation.

- *Never* place tools between panels in a two-speed side slide door, because the interlock is not affected by that action. This action will create a dangerous situation and will not help us gain entry to the car. The insert panel can be readily seen in figure 16–4. The inner panel will move at twice the speed of the outer panel, until they move together completely to a full opening or closure, depending on the direction it is moving.

Fig. 16–4. A two-speed side slide door

The center-opening door has its interlock located at the top of the center line of the door (fig. 16–5).

Fig. 16–5. Center-opening two-speed slide door

This door requires the Halligan to be placed at the top center of the door opening. This would usually be the location of the interlock. At times, if an interlock has been forced a number of times, it may require the repair mechanic to move it further along the top of the door structure, to get a good "bite" when fastening the bolts in to hold it. As with any of the other doors, tools placed down at the bottom, between the two-speed panels, for example, serve no purpose and only make a bad incident into a miserable one.

Other doors include the following:

- Swing door (residential)

- Pull/slide door (residential)

- Biparting door (freight)

Fig. 16–6. Swing door

Fig. 16–7. Pull/slide floor door

Swing Doors

The swing door unfortunately has been associated with many terrible accidents due to the gap between the back of the floor door and the front of the car gate (fig. 16–6). See chapter 10, Residential and Special Elevators, for an extensive section devoted to this problem and its resolution. As a firefighter responding to this type of entrapment, the floor door can be forced with tools at the top right corner. This is where the interlock housing is located. A sister to this door is the pull/slide door for the very same type of installation (fig. 16–7). The difference is that instead of swinging out, the door pulls to the side. Both the pull/slide and swing doors usually have a gated car door, as seen in figure 16–8. None of these doors should present a problem to a fire company that is well trained on its forcible entry equipment. A point to remember is this: You will nearly always find the interlock at final closure contact point.

Fig. 16–8. Slide car gate

Fig. 16–9. Biparting freight elevator door

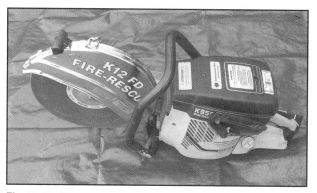

Fig 16–10. Power saw

Rabbit Tool

Generally, the Halligan tool does the job when needed, but to make things a lot easier, the rabbit tool has come along. It, and all of the other similar tools, has replaced bulk and brawn with solid, well-positioned, constant pressure (fig. 16–11).

Biparting Doors

When confronted with a rescue situation involving biparting doors, as shown in figure 16–9, remember that they are different. They open away from each other, and are hung on a wire rope and pulley arrangement that supports both doors. To force entry, when necessary, the force should be applied equally and simultaneously to both the left and right sides of the door. The tools should be placed close to the outer edge, between the rubber astragal that separates the metal door sections, and great care should be taken to keep the door section rising at the same rate. If not, the door will rise high on one side and possibly fall out of the pulley and wire rope arrangement, falling back onto or into the elevator. An alternate method to consider is to use a power saw and cut the skin of the door, then reach in and trip the release on the inside of the floor door (fig.16–10). The biparting door is one of the more difficult doors for forced entry.

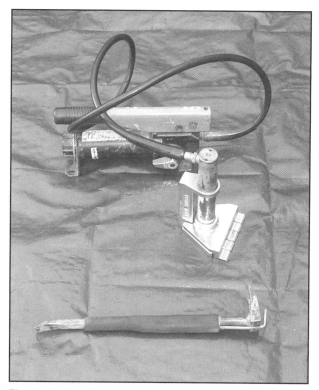

Fig. 16–11. Rabbit tool and J bar

The world of the small compact pressure-producing units has thankfully replaced the forcible entry tools of the not-so-distant past. The J bar helps provide the initial bite, or purchase, as some members call it, which enables the teeth of the rabbit tool itself to be inserted. Pressure is accomplished by pumping on the handle until the obstruction (the keeper) is overcome. Pressure is released by opening the relief valve on the unit.

Air Bag System

Most if not all fire departments carry air bag systems to provide lift for heavy extrication incidents (fig. 16–12).

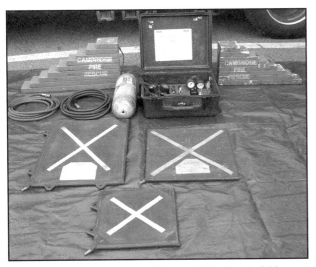

Fig. 16–12. Air bag system with wood blocking/cribbing

These bags can also be used in elevator door forcible entry operations, by using the smaller bag as the pressure producing object. The small bag can be inserted into the opening provided by the J bar or Halligan tool, and pressure can then be produced by slowly filling the bag with air. The wood "cribbing" is necessary to provide a level surface to rest the bag on, to protect the bag from sharp surfaces, and to fill in void spaces.

If the incident turns out to be a crush injury, with a mechanic's or a child's life depending on your forcible entry skills, then the whole chest of tools is available. If after the floor door has been forced, it is discovered that there is a need to repel down to the car top, full harness protection is necessary (fig. 16–13). Only members who

are fully trained should attempt this type of operation. After members have safely arrived at the car top, they should engage the car top inspection station emergency stop switch and turn on the light provided on that unit. Access to the car can now be obtained by opening the fasteners holding the top of car hatch closed. As was shown in earlier chapters, the light baffle is easily removed, and rescuers can be lowered down to provide any necessary assistance (fig. 16–14).

Fig. 16–13. Harness/rope/helmet protection

Fig.16–14. Open hoistway

There are times when the incident itself has happened outside the elevator. Elevator mechanics and helpers have been seriously injured and killed by moving elevator equipment that somehow caught them while in the hoistway. Sometimes, as a regular part of their duties, they are required to work around moving elevators, performing inspections and routine maintenance. Firefighters are called for a man "stuck in the elevator," and the nightmare situation begins. This also has included children and others caught by defective equipment or while playing near the machinery.

Another group is made up of adolescents or young adults, testing themselves against the machine, and unfortunately they have lost (surfing, i.e., riding the top of moving elevators). Whatever the age or occupation, when a person is struck by an elevator or one of its components, it is usually an entrapment that will test every skill that firefighters possess.

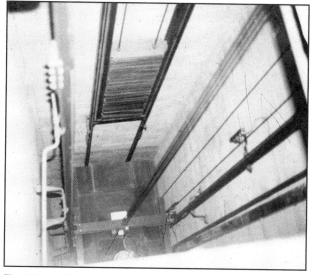

Fig. 16–15. Elevator in hoistway

An important point to keep in mind is that before the firefighter sets foot onto the top of the car, it must be confirmed that the hoist ropes are taut. During a crush injury, there may be some slack rope sitting on the top of the car, or coiled on the floor in the machine room. This slack rope must be taken up by an elevator mechanic before anyone steps onto the car. There is always a danger that the car could be freed from the jam up and would fall down the length of that slack rope. The elevator must be secured before anyone enters the hoistway, and lockout/tagout performed (fig. 16–15).

It cannot be emphasized enough that elevator personnel must be on scene during any extrication attempt involving a person inside the hoistway.

After establishing the condition of the person crushed by the car or counterweight, it must be decided how the person will be freed. Hand tools and young strong bodies can provide some relief for a hand or foot, but a whole body or limb require more drastic steps. With the assistance of the elevator personnel, there are steps that can be taken to provide more room, so that the person can be removed to safety, and they include the following measures:

- Secure the elevator by fastening it on both sides with slings from come-along type units and lifting straps rated for this heavyweight work. They can be fastened onto the rail guide brackets *above* the car and attached to the crosshead beam area.

- To gain a small amount of movement, insert long steel pry bars into the space on either side of the entrapment. It is a fact that elevators crush very effectively, and the entrapped part may be freed readily.

- If that did not provide the needed movement, then the mechanic may want to remove the guide wheels on the side affected to provide a limited amount of movement. This may allow the car to be moved slightly back in the hoistway, ideally enough to provide relief (fig. 16–16).

Fig. 16–16. Roller guide set

The next step would be to cut the guide rail brackets above and below the elevator, allowing the car to be pushed back or forward slightly in the hoistway, allowing room for victim removal.

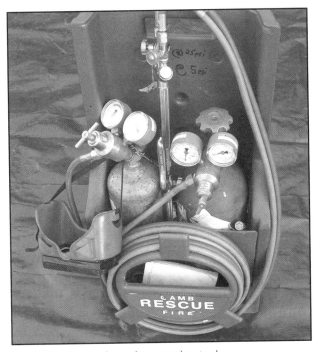

Fig. 16–17. Oxygen/acetylene cutting tool

- Finally, after assuring that the car is absolutely secure, and the trapped person is secured, a mechanic may request that the rails be cut above and below the car on the entrapment side (fig. 16–17). The car is still attached to the guide rail on the other side, allowing it to be swung similar to a gate. Ideally, that would provide the movement to allow the extrication of the victim.

- Finally, when all else has failed, mechanics and firefighters working as a team have brought about amazing resolutions of situations by the ingenuity of those involved.

The crush injury response requires coordination of many forces, including fire department, elevator companies, city or state inspectors, EMS, police, and building management.

Summary

The use of forcible entry tools and equipment has had a bad reputation among those who are not firefighters. What others from outside of our profession fail to realize is that there are two kinds of damage brought about by forcible entry:

- Necessary damage

- Unnecessary damage

When we as firefighters have a job to do, we choose between the two. If we can wait, then we wait. If not, then we do not. The real decision is not whether we will do damage, but how to do it in the least destructive way possible.

Once the decision is made to force entry, then the decision should be acted on swiftly and with direction as to what is the desired effect and how to get it done best. An alternative plan should be considered at the same time in event the path that we chose proves to be extremely difficult or impossible to achieve successfully.

We have reviewed the various doors that will present a challenge to our entry, and the precautions and mandates (lockout/tagout) that we must follow.

As a situation that appeared to be a stalled elevator call turns into a crush or entrapment injury, our skills and ability to adjust to an increasingly demanding incident will be tested. The stalled elevator call handled by a limited response is now a larger incident, with a safety officer and other staff assignments filling in the now expanding needs of the incident. The selection of heavy rescue tools and appliances necessary for a successful outcome may require a mutual aid response from other communities, as most departments do not carry all that may be needed.

The most important tool that you will need for this type of incident is one that you should have requested on arrival, at what was thought to be a stalled elevator. The *elevator mechanic* has all of the needed skills and knowledge to assist with the demands of this incident. Without the mechanic there, it will be similar to responding to a building fire without our best friends on the truck—the irons.

Review Questions

1. What tools does the Halligan replace?

2. List the operating sequence when using air bags.

3. What is the most important tool available to you?

4. Define and explain the two types of damage done by a fire department.

5. Name and describe the most difficult type of floor door encountered.

Field Exercise

Create a database of the various types of forcible entry equipment available to your department for a crush injury incident. Catalog them along with whatever other superior tools may be available in your mutual aid response district.

Chapter 17

Managing Passenger Entrapment Incidents

Introduction

In a typical elevator incident, an elevator has stalled with passengers aboard. There is no emergency, just one or more passengers temporarily "locked in a box." This disruption to their routine is an inconvenience. It has delayed their schedule of activities. Yet, to them, it's more than an inconvenience or imposition. The longer they remain inside the "box," the more impatient or anxious they may become. The firefighter's mission is to calm the passengers and do what is necessary to safely evacuate them from the stalled car.

Preferably, firefighters should work together with the elevator mechanic to evacuate passengers. If the elevator mechanic is nearby, and the incident is not an emergency, firefighters should await the mechanic's arrival. The elevator mechanic usually can safely move an elevator to allow trapped passengers to walk out through the door at the landing. However, the elevator mechanic's response time and the nature of the elevator incident may require firefighters to implement rescue guidelines before the elevator mechanic arrives.

As noted throughout the book, safety is of paramount concern. The guidelines discussed in the chapter are built on comprehensive classroom and hands-on training,

written elevator rescue guidelines, and strong attention to safety for both rescuers and passengers. A key player in firefighter preparedness is the elevator mechanic.

Firefighters should not attempt to move an elevator car except in the situation where they use Phase I of firefighter's service (discussed later in this chapter.) Otherwise, if the stalled elevator requires movement by other means to safely evacuate passengers, the elevator mechanic must do that, but *no one else*.

In a typical elevator incident with passenger entrapment, the elevator car is found stalled at or near the landing, within three feet of the landing, or more than three feet from the landing. Most of these types of incidents are found to be routine in nature; that is, there is no real emergency, just passenger inconvenience.

Sometimes, however, firefighters arrive to find a passenger sick or injured in the stalled elevator car. This situation is a medical emergency and commands immediate action by firefighters to quickly, though safely, gain entry to provide emergency care and evacuate the distressed passenger.

The chapter is divided into two main sections: Passenger Removal: Nonemergency Guidelines and Passenger Removal: Emergency Guidelines. Generally, the main differences between these two types of incidents are the following:

- The nonemergency incident usually does not carry a sense of urgency. Unless the waiting time for an elevator mechanic is determined to be excessive, firefighters should continue to reassure the passengers and await arrival of the elevator mechanic.

- In the case of an emergency incident, time is a crucial factor. Therefore, the urgency of rescue operations often dictates that firefighters must begin their action plan before the elevator mechanic is available.

- In an emergency situation, forcible entry often is necessary to facilitate rapid entry into the elevator car. Yet, if practical and not too time-consuming, firefighters should consider using a hoistway door key, elevator poling tool, or both to unlock a hoistway door(s), rather than using forcible entry.

Safety Principles

Irrespective of the position of the elevator car or the nature of the incident, before attempting to remove passengers, firefighters must ensure that the main power to the elevator(s) is removed and lockout/tagout procedures are followed. Assign a firefighter to remain outside the machine room to prevent unauthorized entry.

Firefighters also must take measures to protect themselves as well as the passengers. The position of the stalled elevator may require that firefighters use fall protection (safety harnesses and ropes), and passengers wear a hardhat and use a safety belt (harness) with rope.

When incident conditions require that firefighters work atop an electric traction elevator, they should use the crosshead (the beam to which the hoistway ropes are attached to the elevator car) as a handhold or otherwise support. Do *not* grab hold of or lean on any of the elevator's hoisting ropes. Depending on how they are arranged, the ropes can move with any

movement of the elevator car. If a firefighter's hand or loose clothing is caught between a rope and the sheave, crushing traumatic injury can result. This should not be a problem so long as firefighters remember to shut down the mainline power to the stalled elevator as well as other elevators as a precautionary measure.

In addition to the safety principles noted earlier, firefighters also should assign a safety officer to oversee rescue operations. Before beginning operations, the incident commander should discuss the action plan, identify hazards, and review other applicable safety measures with other members of the team.

Locating the Stalled Elevator Car

After entering the building, one of the first actions that firefighters should take is to locate the stalled elevator. One quick way is to check the hall position indicator above each elevator entrance at the lobby level, if provided. It might indicate where the elevator car is located.

Another way is to contact passengers using the telephone or intercom system directly linked to the stalled elevator car. Passengers might have a sense of where the car is located. If the previous methods are not helpful, then firefighters should open a hoistway door at the lowest landing where a keyhole is provided to shine a light up or down the hoistway as necessary to find the car. Counting the number of hoistway doors to the stalled elevator car can determine the closest accessible landing.

Establish Communication and Reassure Passengers

On reaching the floor nearest to the stalled elevator car, firefighters should reestablish communications with the passengers. This is especially important if power is lost to the building, and there is no emergency generator. In this case, firefighters should use a light to help illuminate not only the inside of the stalled elevator car but also the rescue area. The presence of light should be a welcome sight for the passengers who would otherwise be in a darkened car.

Firefighters should note that many elevators are provided with an emergency in-car lighting system. It has a four-hour battery pack located inside the car operating panel (COP) or gang station. The battery pack also supplies power for one hour to the alarm bell.

Gather Information From Passengers

Firefighters should ask the passengers the following questions to help determine which action plan is necessary to manage the incident:

- How many persons are in the elevator car?

- Are any of the passengers ill or injured?

- Are the lights on in the elevator car?

- Is the emergency stop switch, if provided, set in the *stop* position?

- How long has the elevator been stalled?

Passenger Removal: Nonemergency Guidelines

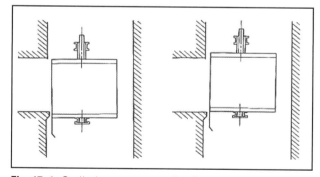

Fig. 17–1. Stalled car at or near landing

An elevator may be found stalled at or near a landing (fig. 17-1). Before discussing guidelines for safely removing passengers from an elevator that is stalled in this way, or where the elevator is either within three feet or more than three feet from a landing, we want to mention the potential benefit of using Phase I operations. Placing the elevator on Phase I may override

the problem that caused the elevator to stall, and allow the elevator to return to the lobby level and open its doors, allowing passengers to exit. However, if any of the safety circuits for that elevator system are open because of an open hoistway door or for other reason, the elevator brake will be applied to the drum, causing the elevator not to move.

A similar situation would result for a hydraulic elevator. In this case, an open safety circuit would cause power to be removed from the hydraulic pump that controls movement of the elevator. Until the problem is corrected the elevator will remain where it is. However, this should not be confused with power down.

Use of Phase I operation of firefighters' emergency operation

To use Phase I operation, firefighters must do the following:

- Ensure the mainline disconnect switch to the stalled elevator is in the *on* (closed) position.

- Ensure that the emergency stop switch inside the stalled elevator car (if provided) is in the *run* position, not the *stop* position.

- Inform the passengers that efforts are underway to move the elevator car to the lobby landing to allow them to leave the car.

- Tell the passengers to remain away from the doors until the elevator car reaches the lobby and opens its doors.

- Use the firefighters' emergency operations key, not to be confused with the hoistway door key, to set the Phase I switch in the *on* position.

If activation of the Phase I operation does not result in recalling the elevator to the lobby level, then firefighters should return the Phase I switch to the *off* position. After doing this, firefighters must power down and use lockout/tagout procedures for the stalled elevator and other elevator(s) as necessary. Keeping the passengers informed about the progress of rescue operations and continuing to reassure them should help passengers maintain their patience and calmness.

Power failure

If the reason for the stalled elevator(s) is because of a power failure in the area, many high-rise buildings are required to have an emergency power generator capable of providing power to all elevators. Under generator operation, power automatically is transferred from one car to another in sequence. Most systems in a large buildings will have all of the cars down and waiting for us in the lobby when we arrive. If a car is stalled, it will skip to the next one after a prescribed wait period. However, if the emergency or standby power system is not capable of operating all elevators simultaneously, then a selection switch is required to permit the selection of the elevator(s) to operate on the emergency or standby power system.

To transfer emergency power to the elevators, it is necessary to activate the elevator emergency recall by using the firefighter's keyed switch capture station. To bring the elevator cars to the main floor, in most cases only one at a time, firefighters must place the system in Phase I and locate and operate the manual elevator standby power selection switch.

Using elevator car power door operator

If efforts are unsuccessful in opening the doors either from inside the elevator car or outside the car at the landing, firefighters should consider using the car door operator mechanism located atop the car. To do this, firefighters must gain access to the hoistway from the landing directly above the stalled elevator car by opening the hoistway door. Using a hoistway door key or a poling tool usually is effective in opening the hoistway door.

After gaining access to the hoistway and depending on the distance from the landing above to the top of the elevator car, a firefighter either must climb down a ladder (that is placed by firefighters) or simply step down onto the car top. Before getting on top of the elevator car, the firefighter should identify potential hazards and a safe location from which to work. On reaching the car, the firefighter should set the emergency stop switch in the stop position. Another precaution is to set the inspection switch to the *insp.* position (fig. 17–2). However, on modern elevator installations, the inspection switch requires a key to operate it. In this situation, firefighters should just rely on the emergency stop switch as an added safety measure.

The firefighter then stands near the car door power operator. (The operator is a motor-driven device mounted on the elevator car that opens and closes the car and landing doors simultaneously.) Placing one foot on the belted pulley of the car door power operator, the firefighter rotates the pulley as necessary, thus

opening the car and hoistway doors together. However, depending on location of the elevator car, and if the elevator car door is equipped with a door restrictor, the door may not open.

A firefighter enters the stalled elevator car, and, if available, sets the emergency stop switch to the stop position. Then the firefighter helps the passengers out of the car, one at a time. If the car is not level with the landing, the firefighter should ensure that the passengers do not trip or fall as they leave the car (fig. 17–3).

Fig. 17–2. Car top controls, including emergency stop switch and inspection switch

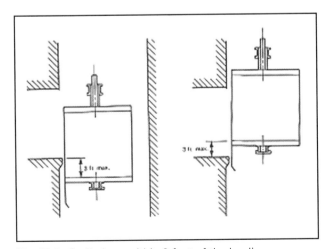

Fig. 17–3. Stalled car within 3 feet of the landing

As with all rescue operations, firefighters must power down the stalled elevator and use lockout/tagout procedures before attempting to open a hoistway door. If the stalled elevator is in a single-car hoistway, a firefighter can use a hoistway door unlocking device key to unlock the hoistway door at the landing where the car is located. However, if the car is in a multiple-car hoistway and a keyhole is not provided, firefighters may be able to open the hoistway door from the adjacent elevator using a poling tool. (See chapter 14, Poling Guidelines, for details.) If there is an elevator on the other side of the stalled elevator that is still in service, firefighters must shut it down for safety reasons before poling across. After bringing the rescue-assist elevator to the nearest landing to the stalled elevator, the mainline power is removed from the rescue-assist elevator, and the emergency stop switch inside the rescue-assist elevator car is set to the stop position. However, an in-car stop switch might not be accessible in some elevators. This is because of a change in the ASME 17.1 Code in 1987/1989 that requires a key-operated stop switch, accessible by the elevator mechanic only. In any case, a firefighter then extends a poling tool between the elevator car and the front hoistway wall to reach the hoistway door drive roller to unlock the hoistway door.

Another firefighter opens the hoistway door, enters the stalled car, and sets the emergency switch in the stop position. If the car is below the landing, then firefighters must place a short ladder into the elevator car to facilitate safe removal of each passenger. Where the floor of the car is above the landing, firefighters should place a short ladder against the floor of the car to facilitate passenger removal. As an added precaution, firefighters also should use a ladder or 42-inch barricade (commercially available) to guard the exposed hoistway opening below the car floor.

Before passengers are assisted from the elevator, be sure to explain the procedure including the rule that one passenger is removed at a time. This will help to avoid passenger confusion, maintain order, and prevent injury. The passengers are assisted from the elevator car by a firefighter inside the car and a firefighter at the landing. It is important that a firefighter maintain a hold on each passenger as he or she climbs down or up the ladder (depending on the position of the elevator car).

After all of the passengers are removed, firefighters should then remove both ladders, close the hoistway door, and check them to ensure they are locked. However, do not restore power to the stalled elevator. Firefighters can retrieve their lockout/tagout equipment after the elevator mechanic arrives and is aware of the details of the incident.

Fig. 17–4. Stalled car more than three feet from the landing

When the stalled elevator car is positioned more than 3 feet above or below a landing, firefighters will need to use the top emergency exit (fig. 17–4). Usually, a ladder is required to reach the top of the stalled elevator car from the landing directly above it. However, sometimes the situation is such that the top of the car is very close to the landing above. In that case, a ladder would not be necessary.

The guidelines presented here will address the situation where a ladder is required to reach the top of the car. Typically, firefighters can open the hoistway door at the landing above by using a hoistway door unlocking device key (if a keyhole is provided) or by poling across from an adjacent elevator (multiple-car hoistway). If both methods are ineffective, then firefighters must consider using forcible entry. The use of forcible entry is considered a last resort and should not be used unless the waiting time for the elevator mechanic is unreasonable for existing conditions, including a situation where a passenger experiences a medical emergency.

After opening the hoistway door, firefighters should identify potential hazards on or near the top of the elevator car. Using a flashlight can facilitate the check. A list of hazards is presented next:

- Car door power operator (moving parts)

- Counterweight assembly

- Open space between the elevator car and hoistway walls

- Smooth floor surface

- Unprotected lightbulb

- Limited work area

- Damage electric wires

- Nearby elevator equipment

- 2:1 roping (the arrangement where ropes move around a sheave attached to the crosshead located on top of some traction elevator cars; the hazard is eliminated once the mainline power is removed from that elevator.)

After implementing standard safety precautions including the use of fall protection, the firefighters lower a ladder, preferably with nonskid feet, to the top of the car. The ladder should be of sufficient length to extend at least three feet above the landing.

A firefighter descends the ladder to the car. Then a second ladder is lowered either by rope or hand to the firefighter for later placement inside the elevator car. This ladder should be of sufficient length to extend at least three feet above top of the elevator car. A second firefighter descends the ladder to assist the first firefighter in receiving and placing of the second ladder. A safety belt (harness) is either lowered or carried by the second firefighter for use in passenger removal.

Firefighters are cautioned that the available surface area on top of the car, types, and location of potential hazards, and even the weight of firefighters are factors that influence how many firefighters should stand on the car top. As a rule of thumb, no more than three firefighters should work at one time on top of the car.

Having too many firefighters on the car top at one time can not only hamper rescue operations but also compromise safety.

After the second ladder is safely set aside, the firefighter locates the top emergency exit, which usually is located on the rear half of the car. Unlocking the exit cover usually requires firefighters to release slide bolts, wing nuts, or sash locks. The ASME 17.1 Code requires that the exit cover be opened only from the top of the car, without the use of special tools. Before this requirement was enacted and elevators were modified, "easy" access to the top of the car from inside the elevator car had allowed some individuals to get atop the elevator car and wait to surprise passengers with the intent to harm or rob them. Also, inside access allowed individuals to "play" atop the elevator car with injurious or fatal consequences. The concerns for safety and liability are factors that led to changes in the ASME 17.1 Code.

Though most top emergency exit panels are hinged, some panels are completely detachable, but with an attached chain secured to the car top. Therefore, firefighters must exercise caution to prevent the detachable cover from posing a fall hazard. However, if the chain is not fastened or is missing, the firefighter should place the cover in a safe location atop the car or pass the cover to firefighters on the landing above.

Once the top emergency exit is open, firefighters may find that the false ceiling, if provided, prevents free access to the inside of the elevator car. In this case, a firefighter should slide the panel to one side, or remove it from the car and pass it up to firefighters at the landing above. The ASME 17.1 Code requires that the opening be free of obstructions.

After ensuring free access to the car, a firefighter tells the passengers to move away from the opening. Then the ladder is lowered through the top emergency exit and positioned between the elevator car floor and car top. The ladder should extend at least three feet above the car top (fig. 17–5).

While one firefighter holds the second ladder, the other firefighter descends the ladder to the car floor. A safety belt (harness) and lifeline is either carried by the second firefighter or is passed down after the firefighter is inside the elevator car. Once inside the car, the firefighter sets the emergency stop switch to the stop position.

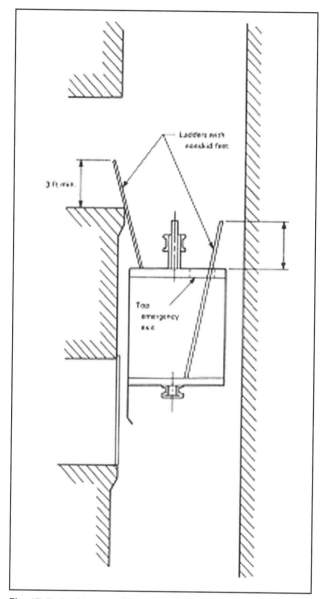

Fig. 17–5. Ladders positioned inside the car and on top of the elevator car

The safety belt and lifeline are secured to a passenger, and the rope is held taught (by firefighters on the landing above) as the firefighter assists the passenger up the ladder. The firefighter standing on top of the elevator car grabs hold of the passenger and assists the passenger up the ladder to the landing above. Given the steps in the removal process, ongoing communications and coordination is necessary among the firefighters engaged in the actual removal of the passengers.

After all of the passengers have safely reached the landing above, the firefighter located inside the car leaves the emergency stop switch in the *stop* position, climbs the ladder, and removes the ladder and passes it to firefighters on the landing above. Then the cover to the top emergency exit is closed or repositioned, and finally, it is locked.

The two firefighters then do the following:

- Check for any tools used in the rescue operations.

- Reset the emergency stop and inspection switches to their *run* positions.

- Climb to the landing above.

- Remove the remaining ladder from the hoistway.

- Close and check the hoistway door to ensure it is locked.

As part of the final check, firefighters should verify that all hoistway doors opened during rescue operations are closed and locked. The extra effort is worth it, both in terms of safety and potential liability.

It is also important that firefighters ensure that any elevator used during rescue operations, *except for the stalled elevator*, is ready to have its main power restored. The decision to restore power to the rescue-assist elevator rests with the incident commander. However, main power must remain off to the stalled elevator, though the lockout/tagout procedure is discontinued. Before leaving the scene, the incident commander should inform the building engineer or other competent management representative of the situation including the status of the elevators.

Passenger Removal: Emergency Guidelines

As mentioned earlier, the urgency of an emergency incident and the likely need for forcible entry separates this type of incident from the nonemergency elevator incidents. The section will focus on the use of forcible entry to gain access to the stalled elevator to attend to the sick or injured passenger. The guidelines used in nonemergency incidents are essentially applicable to emergency incidents. Therefore, for the sake of brevity, we will not repeat them. The focus will be on gaining access to the distressed passenger and providing needed emergency care and removal.

If the distressed passenger or other passenger can communicate with the firefighter, the firefighter should attempt to obtain more information about the distressed passenger's medical condition. The additional information can help the firefighter decide the next course of action, if any, before entry is made into the elevator car. The action may be to provide CPR instructions.

Firefighters are reminded that if an elevator mechanic is present, there should be no need to force entry into the car. If the elevator mechanic is not at the scene, then firefighters should use a hoistway door key or poling tool to quickly unlock a hoistway door, where applicable and practical. Otherwise, firefighters must use forcible entry.

Forcible Entry Techniques

Forcible entry considerations

Before forcing a door, the firefighter should instruct passengers to move to the rear of the car away from the door. Preferably, firefighters should use the tools listed below in the following order:

- Hydraulic forcible entry tool
- Prying bar and flat head axe
- Lifting bag (mini)

To open a hoistway door, firefighters must use the prying tool to apply force as close to the top of the door as possible. This usually is where the interlock is located. *Do not* apply force at the bottom of the door. This dangerous situation ultimately can result in the door falling down the hoistway.

The firefighter should insert the tool between the door and the doorjamb on side-sliding doors and between the doors on center-opening type. Firefighters should not confuse a two-speed, side-sliding door with a center-opening door. The surfaces of the center-opening doors are flush with each other. When forcing a two-speed side-sliding door, the firefighter should apply force at the doorjamb.

After opening the hoistway and car doors, firefighters enter the car, set the emergency stop switch in the stop position, attend to the injured or sick person, and then

remove the passenger to a landing. If other passengers are in the car, firefighters should remove them first to provide more space to care for the distressed passenger.

Once all of the passengers are removed safely from the elevator car, firefighters should follow the relevant precautions and actions consistent with nonemergency elevator incidents.

Rescue from a blind hoistway operations

The ASME 17.1 Code defines a blind hoistway as "the portion of a hoistway where normal entrances are not provided." Blind hoistways are found in tall buildings and buildings that extend deep below the ground. Examples include high-rise office and residential buildings, observation towers, and buildings in which passenger elevators serve underground rail systems (subways). The ASME 17.1 Code also states, "Where an elevator is installed in a single blind hoistway, there shall be installed in the blind portion of the hoistway an emergency door at every third floor, but not more than 36 feet" that conform to certain conditions.

Depending on the distance from the closest emergency door to the top of the stalled elevator car, the use of a ladder to reach the car may be impractical. In this situation, firefighters must carefully evaluate whether it is more prudent to wait for an elevator mechanic than to rappel down the hoistway. We recommend that unless there is an emergency, firefighters should wait for the elevator mechanic.

However, if an emergency exists that clearly indicates that rappelling is the only viable alternative to waiting for an elevator mechanic, the incident commander must only allow firefighters with sufficient rappelling experience to engage in the actual rappelling operations. Preferably, members of a high-angle or technical rescue team should execute this aspect of the rescue mission. The rescuer doing the rappelling either should carry an emergency jump bag or later have one lowered to the car.

Firefighters should comply with the safety principles, applicable precautions, and guidelines discussed earlier in the chapter. After the descending firefighter is safely positioned on top of the elevator car, the objective should be to establish communications with the passengers, assess the situation, and provide emergency care instructions as necessary.

The next step in the rescue operation simply may be to wait for the elevator mechanic to arrive. After the elevator mechanic arrives and confers with the incident commander, the probable remedy is to have the elevator mechanic control the lowering or raising of the elevator car to the nearest landing. Communications and coordination among the firefighter in the elevator car, incident commander, and the elevator mechanic is crucial to an effective and safe rescue outcome.

Documentation

Documentation of all types of incidents is a common and necessary task of firefighters. This also is true for elevator rescue incidents. Although routine elevator incidents (passenger inconvenience) usually are relatively simple to do, emergency incidents will require more detail.

Taking photographs, preparing line sketches, and taking witness testimony are ways to capture essential details of an elevator incident where a victim is injured or killed. The written comments of firefighters who were engaged in rescue operations is a common, yet important, part of the documentation process.

This detailed report will be beneficial not only to the local fire or rescue department for training and other uses but also to the state or local elevator safety authority and the Occupational Safety and Health Administration (OSHA). Elevator incidents that result in serious injury or death often lead to comprehensive after action reports (AAR), review of elevator rescue standard operating guidelines, legal inquiries, or civil lawsuits.

Passenger Entrapment Rescue Operations Review Points

- Assemble and carry elevator tools and equipment into the building.

- Follow the safety principles and rules.

- Locate stalled elevator.

- Use designated telephone or intercom system to contact passengers inside the stalled elevator car.

- Establish direct communication with and reassure the passengers.

- Determine if the elevator incident is an emergency or nonemergency.

- Evaluate situation to determine if it is advisable to wait for the elevator mechanic or to proceed with rescue operations.

- Develop the action plan including safety measures, identifying potential hazards, and designating a safety officer.

- Recognize that a door restrictor, if present, may prevent the use of hoistway door unlocking device keys, poling tools, or other methods described in this chapter and elsewhere in the book.

- Use the expertise and resourcefulness of the local elevator mechanic.

- Implement the applicable passenger-removal guidelines based on the position of the elevator car and the condition of the passengers.

- Use a lifeline to secure passengers during their removal from the car to the landing, where necessary.

- If the lifeline should possibly slide across the corner of the floor where it joins a hoistway wall during use, consider placing a turnout coat on the floor as a buffer.

- Remove all passengers.

- Secure the elevators and recheck all doors to ensure they are locked.

- Restore power to the rescue-assist elevator or other elevator placed out of service during rescue operations except for the stalled elevator, and remove lockout/tagout equipment.

- Do not restore power to the stalled elevator.

- Inform the elevator mechanic or, if the mechanic is absent, the building engineer or other competent member of the building management about the incident.

- Be sure the incident is thoroughly documented after returning to the station.

- Review the incident not only with members involved in the elevator rescue incident but also other members. The point here is to determine if the standard operating guidelines are sufficient or require change, as well as to improve firefighters' elevator rescue skills.

Summary

Shortly after entering the building, firefighters should locate the stalled elevator car and establish communications with the passengers. Firefighters should inform them that they are safe and steps are being taken to evacuate them from the elevator car. Unless otherwise instructed by a firefighter, passengers should stand away from the front of the car.

If the elevator mechanic is nearby and the incident is not an emergency, firefighters should await their arrival. The elevator mechanic usually can safely move an elevator to allow trapped passengers to walk out through the door at the landing. However, elevator response time and the nature of the elevator incident may require firefighters to implement rescue procedures before the elevator mechanic arrives.

Most of the elevator incidents are not emergencies. Firefighters will find the stalled elevator at or near a landing, within three feet of a landing, or more than three feet from a landing. In the last case, the floor of the car may be either above or below a landing. The position will result in a variation in rescue operations for each case.

If a passenger is sick or injured, then firefighters will have a greater sense of urgency to gain entry to the elevator car, as compared to a nonemergency situation. Moreover, using forcible entry is a reasonable action. Yet, firefighters should not abandon the initial use of nonforcing techniques if their use does not result in significant delay.

Because of the different situations in which firefighters can find a stalled elevator and the circumstances surrounding them, firefighters must have written rescue standard operation guidelines (SOGs). Practicing these guidelines with sufficient regularity enables firefighters to acquire proficiency and confidence in their use. Requesting and using the expertise of an elevator mechanic during training and actual elevator incidents is the wise and safe thing to do.

Some buildings have blind hoistways. If the position of the stalled elevator car is too far below the nearest emergency door of a single-car hoistway, the use of a ladder to reach the car will be difficult or impractical. In this case, firefighters must decide whether to wait for the elevator mechanic or rappel to the top of the car. This is a situation that requires careful evaluation. Firefighters must keep the risk of harm in an acceptable balance with the rescue mission.

New elevators are likely to have door restrictors. These restrictors can prevent firefighters from opening an elevator's car door. Firefighters should become familiar with the types of restrictors installed on elevators in their response areas. The local elevator mechanic is an excellent resource to firefighters to help them learn more about the specifics of the door restrictors found locally.

Firefighters must remember that their desire to evacuate passengers from a stalled elevator car must be balanced with a clear need to do so. If the conditions clearly indicate the need to evacuate the passengers, following a systematic approach, based on sound safety principles and elevator rescue SOGs, should result in successful rescues.

The content of the chapter is based on material presented in the ASME A17.1-2004 and A17.4-1999 publications. The diagrams were taken from these publications as well. Permission granted by the American Society of Mechanical Engineers. All rights reserved.

Review Questions

1. When the bottom of a stalled elevator car is located more than three feet above the landing, how should firefighters remove the entrapped passengers?

2. What two safety precautions should firefighters take to prevent movement of a stalled elevator car before removing entrapped passengers?

3. If forcible entry is required to open a hoistway door, where should the prying action be applied, (a) at or near the top of the door or (b) near the bottom?

4. List one method of protecting firefighters from the unprotected space that is sometimes created between the stalled elevator car and the landing below.

5. When placing a ladder to the top of an elevator car from the landing above or inside the car to assist with the removal of passengers, how far should the tip of the ladder extend above the landing or car top?

Field Exercise

Conduct a training session using an elevator building to practice guidelines to simulate removal of passengers through the top emergency exit.

Chapter 18
Managing Pinned-Victim Incidents

Introduction

An elevator incident involving a person pinned or caught between elevator equipment does not occur that often. Yet, when it does, it usually requires a substantial commitment of time and resources of firefighters. How the victim is pinned and the extent of crush or traumatic injuries can challenge the extrication skills and ingenuity of the firefighters.

The chapter shares information to help firefighters gain a better understanding of the risks and challenges attendant to safely manage person pinned in a hoistway. Contacting other firefighters not only in neighboring jurisdictions but also in fire departments such as New York, Los Angeles, Chicago, Atlanta, and Boston can lead to other useful elevator extrication techniques and procedures. We are simply sharing some ways to do the job.

Depending on the circumstances surrounding the incident, firefighters can arrive to find that the person is pinned between any of the following:

- An elevator car and a hoistway wall or counterweight

- The counterweight and a hoistway wall

- The underside of an elevator car and the bottom of the hoistway

- Other elevator equipment

Victim extrication requires the combined resources of firefighters and the elevator mechanic. The firefighters bring field experience, ingenuity, and tools and equipment. The elevator mechanic brings technical expertise and knowledge not only of the entire elevator system but also knowledge of the specific elevator equipment involved in the incident.

One incident that we recall is one in which a person was killed when the elevator car moved upward while he was attempting to do "pull-ups" at the car entrance. When the car moved, it pinned him between the elevator car and the front hoistway wall.

The incident commander conferred with the elevator mechanic as to the best way to remove the victim. They decided to do this by moving the car from the machine room. Before doing this, firefighters secured a rope around the victim's body to prevent it from falling. While using a portable radio to maintain communications with the incident commander, who was located at the landing directly above the elevator car, the elevator mechanic slowly moved the car down a short distance and then paused to confer with the incident commander. The movement still was not enough to completely free the victim's body. The elevator mechanic slowly moved the car again. This time the victim's body was slowly lowered into the elevator car. The remainder of the incident was uneventful.

This example highlighted the importance of coordination between the incident commander and the elevator mechanic. Both worked as a team to safely remove the victim. The movement of the elevator car by the elevator mechanic was all that was necessary to free the victim.

The safe and successful management of the incident was attributed to:

- Power down of the elevator and using lockout/tagout procedures

- The expertise of the on-scene elevator mechanic

- Communication and coordination between the incident commander and the elevator mechanic

- Securing the victim's body to prevent it from falling

- Development and implementation of an effective action plan

The incident could have been much more challenging and labor-intensive to firefighters if the victim had been alive and had a crush-type injury, and if the elevator had signs of serious damage such as slack in hoist ropes.

Under these conditions, firefighters must not enter the hoistway, or stand on top of or enter the elevator car, until the elevator mechanic has assessed any damage and has determined what actions, if any, are necessary to prevent possible movement of the elevator car during rescue operations. The presence of slack in hoist ropes on top of the car, counterweight, or in the machine room is a sign that the elevator car or counterweight could be in danger of suddenly moving if firefighters were to use tools and equipment to free the victim.

Chapter 16, Forcible Entry, discusses recommendations on how to help secure the elevator car to prevent it from possibly moving. When there is slack in hoist ropes, powering down the elevator alone is not enough

to ensure that the car or counterweight won't move. Chapter 16 discusses guidelines for firefighters to consider when deciding what other actions are necessary.

Initial Questions

Shortly after arriving on the scene, the incident commander should seek answers to the following questions.

- Are there are any witnesses to what happened?

- Where is the elevator car located?

- What is the condition of the victim?

- Is there slack in the hoist ropes?

- In what direction was the elevator car traveling?

Emergency Care

With these types of incidents, the victim can and usually does receive either serious or fatal crush injuries. In anticipation of this, paramedics are part of the fire department dispatch. They can assess the victim's condition, establish an intravenous line, and stabilize injuries as much as possible based on existing conditions.

In some situations, the dire condition of the pinned victim requires emergency medical intervention by a hospital's trauma team or emergency department's doctor. The management of this type of incident is similar to extricating a trauma victim pinned in the wreckage of an automobile or truck. However, a major difference is the often precarious position of the victim and the questionable stability of the elevator suspended in the hoistway. Extrication can be a long, arduous task.

Technical Rescue Team

Depending on the particular fire department, the response to and management of elevator emergencies usually rests with crews of the engine company, ladder company, rescue squad, or technical rescue or special operations teams. For most types of elevator incidents, it doesn't really matter what unit firefighters are assigned to. What does matter is how well they are trained in the subject and how well they follow their standard operating guidelines.

In those situations where a person is pinned between an elevator car and a hoistway wall or a similar situation, there is merit in dispatching the technical rescue or special operations team. There is a distinct advantage in having a cadre of rescue specialists assigned to respond to this type of incident elevator. The main reason is that they are highly skilled in the use of an assortment of rescue tools and equipment designed to breach, cut through, pry, lift, and support heavy objects such as an elevator car or breach a hoistway wall or door.

Another reason is that these firefighters are likely to have more training and experience in elevator rescue operations than firefighters assigned to an engine company or truck company. It is because of the very nature of these specialty teams that they lend themselves to being better prepared to handle the special circumstances of elevator rescue operations involving a person who is pinned with crush-type injury.

However, we should not forget that an incident of this type would require more resources than just an engine company, truck company, rescue squad, or specialty team. The needed complement of resources would also include the elevator mechanic, EMS units, other fire department units for staffing, and possibly a hospital's emergency trauma team.

Safety and Operational Points

- The safe and successful management of the incident requires teamwork and coordination between the incident commander and the elevator mechanic.

- Power down and lockout/tagout procedures are fundamental, yet essential, measures taken by firefighters. Also be sure that other nearby elevators in the same hoistway are shut down and lockout/tagout procedures are used. Be sure a firefighter is assigned to guard entrance to the machine room.

- Assigning a safety officer is both important and necessary.

- Slack rope on top of an elevator car or counterweight or in the machine room is an indication that the elevator car or counterweight might suddenly fall.

- Firefighters must not enter the hoistway, or stand on top of or inside the elevator car, until the elevator mechanic says that it is safe to do so.

- If it is necessary that the elevator be moved up or down from controls in the machine room to free the victim, this is the sole responsibility of the elevator mechanic. Firefighters must not attempt this.

- Firefighters should use a rope or other suitable means to prevent the victim from falling during rescue operations.

- Unless it is absolutely necessary and the safety officer authorizes it, the elevator mechanic should not be working alongside firefighters on top of the elevator car or otherwise in the hoistway.

- Loosening or removing the top roller or slide guides of the elevator (electric traction only) can provide for additional lateral movement of the car away from the hoistway wall. The elevator mechanic should supervise this operation.

- Do not loosen or remove the roller or slide guides of a hydraulic elevator. The added movement of the car can place dangerous stresses on the plunger-cylinder system. If the system should fail, it can have catastrophic results, causing the elevator car to fall.

- The victim's medical condition and estimated extrication time might require the response of a hospital's trauma team.

- Firefighters working in or near the hoistway should be wearing head, eye, hand, and foot, and fall protection equipment.

- To maintain efficiency, eliminate confusion, and reduce the risk of injury, allow only enough personnel to work near or in the hoistway, including top of the elevator car.

- As part of emergency planning and training activities, confer with the local elevator service mechanic to identify ways to temporarily secure an elevator with slack in hoist ropes.

- If an air bag or other equipment is used to move an elevator car away from a hoistway wall, be careful not to cause collapse of the wall.

- If a victim is pinned underneath the bottom of an elevator car, consider using air bags and wood cribbing to lift and support the car.

- Members of collapse rescue and urban search and rescue teams are very skilled in the use of shoring equipment and cribbing materials. Their expertise can be helpful during elevator extrication incidents.

- See chapter 16, Forcible Entry, for additional rescue considerations.

- Firefighters should weigh the risk of using an oxyacetylene torch or other equipment that could be a source of ignition against the benefit of its use.

- Whenever it is required that firefighters work from an exposed hoistway, be sure that a firefighter is assigned to guard the opening and also use a barricade, if applicable. A door wedge tool is useful in holding the hoistway door open. (Its use is in addition to the assigned firefighter and use of a barricade.)

Summary

Managing an incident involving a person pinned between elevator equipment and a hoistway wall can be the most challenging of the incident types discussed in the book. The pinned-victim incident is a combination of a confined space, extrication, and crush injury incidents. Additionally, the fall potential and the possible movement of the elevator car or counterweight increase the risk of injury to firefighters.

Addressing the four safety principles emphasized in the book is fundamental to any elevator rescue incident (see chapter 11). However, in the situation where the elevator car or counterweight might suddenly move, the elevator mechanic should inspect the elevator car from a landing to determine if it is safe for firefighters to enter and work in the hoistway.

A safe and successful rescue operation also requires teamwork between the elevator mechanic and the incident commander. In some incidents, the elevator mechanic can free the victim by using equipment located in the machine room. It is more likely, however, that the rescue scenario would require firefighters to use rescue tools and equipment to create enough space to free the victim. The situation may require coordinated actions by firefighters and the elevator mechanic.

Before any efforts can be made to free the victim, firefighters should secure a rope to the victim to prevent him or her from falling down the hoistway. Moreover, it is important to take other measures to protect the victim when firefighters are working with tools and equipment directly above. If it is necessary to breach a hoistway wall, firefighters should ensure that the action does not result in debris falling and possibly striking the victim.

If the victim has crush-injury trauma and the estimated extrication time is long, the incident commander should consider the benefit of requesting a hospital's trauma team. In the interest of the victim's welfare, the incident commander should make this request early in the incident based on anticipated need rather than obvious need. Time is crucial in these types of incidents.

Fortunately, a pinned-victim elevator rescue incident is not very common. Yet, it is one that requires the most time, commitment, and resourcefulness of firefighters and the elevator mechanic. Through training and elevator-site visits, firefighters can gain and maintain the knowledge necessary to safely and successfully manage this often challenging incident.

Review Questions

1. It may be necessary to lower the elevator car using controls in the machine room to free the person. Who should do this?

2. Before beginning rescue operations to free a pinned victim, you notice slack in the hoist ropes. What should this indicate to you, and what should you do about it?

3. List three ways in which firefighters could find a victim pinned between elevator equipment.

4. List the minimum personal protective equipment firefighters should wear or use during efforts to free a pinned victim.

Field Exercise

Schedule a meeting with the local elevator mechanic to discuss challenges, concerns, safety precautions, and rescue plan required to safely manage a pinned person incident. Subsequent to the meeting, visit a building with the elevator mechanic to get an up-close look at the elevator equipment and to discuss details of the earlier meeting.

Chapter 19
Fall from Heights: Victim Rescue

Introduction

Managing a fall-from-heights incident usually is a straightforward operation for firefighters. Whether a person falls off scaffolding, a roof, or another dangerous, elevated place, the general approach is the same: Assess the scene for hazards, formulate a plan, address safety precautions, and provide emergency care and transport for the victim. This response philosophy also applies to the situation in which a victim falls down a hoistway.

A fall down a hoistway can result from a person standing or working too close to an unprotected hoistway opening, a mischievous teenager "playing" atop an elevator car, or an impatient passenger who attempts to escape a stalled elevator before firefighters arrive.

The major challenge to firefighters will not be providing emergency care to the victim lying at the bottom of the hoistway, but rather ensuring that the rescue work environment is safe for entry and rescue operations. This requires firefighters to power down all elevators in that hoistway and also follow lockout/tagout procedures. It's not only the elevator located directly above the victim that is a big concern but also the ones that might be located on either side of it. During rescue operations, a firefighter might unknowingly move into the path of an approaching nearby elevator car or its counterweight without realizing it. Removing the power to the elevators first eliminates this concern.

Rescue Approach

Once the power down and lockout/tagout issues are addressed, then firefighters can do a quick survey of the pit area and the hoistway space directly above. Doing the survey from the landing or pit entrance is necessary from the safety perspective.

Remember, if a pit access door is provided, it will require a key to gain entry. Be sure you have the key to minimize delay. Otherwise, you will need to use forcible entry.

After opening the pit access door, a firefighter should set the pit emergency stop switch, if provided, in the stop position. If there is no pit access door, then firefighters must gain access by opening the hoistway door at the landing immediately above. A firefighter can do this by using a hoistway door unlocking device key, poling tool, or, as a last resort, forcible entry tool. In any case, firefighters should use fall protection, especially if the distance from the landing to the bottom of the pit is six feet or more.

After the hazard survey is completed and the safety officer declares that it is safe for firefighters to enter the pit area, entry is made. If access is made from a landing, place a ladder into the hoistway to the bottom of the pit, with at least three feet of the ladder extending above the landing. After doing this, the safety officer should ensure that just enough firefighters enter the pit area to render emergency care and safely remove the victim. Allowing too many firefighters in the rescue area can lead to inefficiency and an increased risk of harm.

While the hazard survey and power down operations are underway, other firefighters should check with witnesses, if any, to determine the circumstances surrounding the fall, including where the fall originated and what happened. They should check the upper floors to ensure there are no unprotected hoistway openings, such as an opened hoistway door.

Safety Measures

When planning for this type of incident, don't forget these points:

- Develop an action plan, maintain appropriate and ongoing attention to safety, and provide timely emergency care and evacuation.

- Treat the incident like a "fall from heights," but be mindful that an elevator car and (possibly) counterweights are hanging above your head!

- Consider the possible gradual or sudden movement of a hydraulic elevator if one is present.

- Anticipate the need for a ladder.

- Limit the number of firefighters working in the rescue area.

- Assign a firefighter to do power down and lockout/tagout not only for the elevator immediately above the victim but also for nearby elevators. Keep the firefighter outside the machine to guard entry.

- Gain access to the elevator pit by using the pit access door, if provided, or by opening the hoistway door at the lowest landing using a hoistway door key, poling tool, or forcible entry tool.

- Before entering the hoistway, confirm that the main power (power down) to elevators is off and the lockout/tagout procedures are followed.

- Use a light to scan the hoistway for potential hazards.

- Set the pit emergency stop switch in the stop position.

Summary

Depending on the height, a fall down a hoistway can be fatal to the victim. Although such a fall is generally characterized as a fall from heights, it can provide additional challenge and risk to firefighters. This is because of the confined space of the hoistway and the presence of an elevator car(s) and other equipment in the hoistway.

Providing emergency care to the victim is not the challenge for firefighters. They respond to trauma-related incidents on a regular basis. The challenge is ensuring that the rescue environment is rendered reasonably safe for firefighters to enter and work.

As with other types of elevator incidents, firefighters must ensure that power down and lockout/tagout procedures are accomplished *before* they enter the pit. Moreover, firefighters must check the area immediately surrounding and above the victim for potential hazards. A rescue plan that is built upon the four safety principles discussed in chapter 11, Elevator Safety, will not only facilitate rescue operations but also provide for the safety of firefighters and the victim. Remember, the number-one priority is to ensure that the main power to each elevator is set in the off position. Failing to do this can result in an elevator car or counterweights moving quietly down the hoistway and striking firefighters with deadly consequences.

Review Questions

1. What is the major challenge to firefighters at the scene of a fall from heights incident?

2. In some instances, access to the pit will be through a pit access door. After opening the door and before entering the pit, what two actions should you take?

3. When a ladder is placed into the hoistway to the floor of the pit from a landing, at least how far should the ladder extend above the landing?

4. Fall protection equipment should be worn by firefighters, especially if the distance from the landing to the pit is how deep?

Field Exercise

Schedule a building visit to view the pit, hoistway, and equipment from inside the hoistway. Be sure you have permission from the property manager or building engineering, and have an elevator mechanic present to ensure a safe visit. Take pictures for use in future training sessions.

Chapter 20
Managing Elevator Fires

Introduction

Other than elevator rescue-related incidents, fire-fighters also respond to reported fires in elevator machine rooms, hoistways, and elevator cars. The severity of these types of incidents ranges from smoldering trash, to an overheated motor, to an electrical fire in the machine room, to a raging fire inside an elevator car. Most of the incidents are minor and require little or no action by firefighters. However, there are times when a rapidly developing fire endangers passengers in elevators, other building occupants, and even firefighters.

Don't let the minor nature of most elevator fire incidents lull you into a state of complacency that causes you to forget to power down the elevator(s) and comply with other safety principles and practices. The risk of harm is often not directly associated with extinguishing the fire but rather the hazards surrounding it.

This chapter focuses on the types of incidents, potential hazards, and safety considerations. We limit our fire-extinguishment discussion to the point of fire origin, not the entire scope of activities attendant to firefighting operations in buildings. That is left to the local firefighters, consistent with their standard operating guidelines (SOGs).

With the exception of a fire in a machine room that is located outside of the hoistway, all other elevator fire incidents occur in the hoistway. This enclosed environment poses inherent hazards and concerns to firefighters summoned to extinguish the fire and otherwise manage the incident.

Hazards

Although we have mentioned most of these hazards in earlier chapters, we think it is important to list them again. Developing a strong appreciation and awareness of these hazards will help firefighters to not only follow appropriate safety measures to protect themselves against these hazards but also protect elevator passengers and other building occupants.

- Fall from heights (exposed hoistway)

- Bare electrical wiring and equipment (shock hazard)

- Hot metal surfaces and exposed lightbulb (burn injury)

- Movement of elevator car and equipment (crush injury)

- Spread of fire, smoke, and toxic gases (inhalation injury)

Elevator Car

There are two fire situations that firefighters can encounter. The most common and usually minor situations are an overheated car door operator motor, shorted fan, or light fixture. The other one is not nearly as common, yet it is one that can result in serious fire consequences and create a high risk of danger to firefighters, elevator passengers, and other building occupants. It is where a serious fire develops in or spreads to the inside of an elevator car.

In one such fire, building custodial staff used an elevator to remove trash floor by floor in a multistory apartment building. Unfortunately, on one occasion, a discarded smoking material ignited trash in an elevator that resulted in a multiple-alarm fire. Though the elevator car remained at the floor, smoke rapidly spread throughout the building, causing some occupants to jump from windows (fig. 20–1).

A very hot fire could cause any number of serious problems with the elevator doors and suspension. Firefighters should be aware that, in worst-case scenarios, it's possible that a car could actually drop some distance before it is stopped on its safeties. During testing, the entire car is not subjected to fire, so there is no telling exactly what might happen if the car is totally involved in fire. Note: In the circumstance where an elevator's suspension ropes and governor rope fail concurrently, the elevator safeties will not set. This condition would allow the elevator car to fall.

Overheated motor or shorted fixture

The telltale signs of this type of incident are its characteristic "electrical" odor and smoke (fig. 20–2). To successfully manage this incident, firefighters must remove passengers from the elevator and other elevators if endangered by the smoke, power down and lockout/tagout the problem elevator, and also remove power from the 110-volt circuit that controls power to the car light and fan. Remember, the main power to an elevator is separate from the power supply to the car's light and fan. So, if you just turn off the main power to the elevator, there is still power to the light and fan.

Fig. 20–1. This picture was taken in another building in the same apartment complex, one day after the fire.

Fig. 20–2. The car door operator motor can be a source of smoke or odor should it overheat.

Depending on existing conditions, firefighters might simply let the motor cool on its own or use a suitable fire extinguisher, such as an ABC type. In either case, after checking to ensure there is no longer a problem from a fire department perspective, firefighters should not restore mainline power to the elevator or the separate power to the car's light and fan. However, firefighters can restore power to the other elevators that may have been shut down unless there are tactical reasons for not doing so.

Fire inside the elevator car

Extinguishing a fire in an elevator car is like putting out a fire in a walk-in closet. About the same flow rate and quantity of water is needed to extinguish the fire. Yet, there is one major caveat that separates both situations. The burning closet is secured to the floor, whereas the burning elevator car is like a container, precariously hanging high in the hoistway, waiting to fall!

A raging fire inside an elevator car can quickly cause warping of its hoistway door in the opened or closed position and also possibly result in the elevator car running down. The car may drop a short distance before its safeties set. That is not true, however, if both the governor rope and hoist ropes separated. In this situation, the elevator can fall.

The burning elevator car may present one of the following situations when firefighters arrive to battle the fire:

- The elevator car is still at the landing but may run or otherwise move down depending on conditions at a moment's notice.

- The elevator car has run or otherwise moved down, thus exposing the hoistway.

- The elevator car either remains at the floor or runs or otherwise moves down while firefighters are present, leaving the hoistway door warped in the open position.

Knowing that the fire involves an elevator car is one of the most important pieces of information for firefighters to know at the outset of fire suppression efforts. Knowing this, firefighters would not get too close to or attempt entry into the elevator car. However, if the burning elevator car is too obscured by flames and dense smoke, and the firefighters don't have previous knowledge of the existence of an elevator in that location on the floor, they may mistake if for the entrance to an apartment unit. And, as such, firefighters are likely to advance their attack line into the "apartment." The result can be a near-fatal or fatal fall down the hoistway.

This dangerous scenario underscores the importance of firefighters conducting surveys of elevator buildings. When gathering information about elevators for use in elevator rescue incidents, firefighters should also consider the typical floor layout, including any unusual arrangement of elevators. Absent prior knowledge of the floor layout through a building survey or check of a typical floor, unsuspecting firefighters would reasonably assume that under fire conditions a fire emanating from one end of the hallway would likely be an apartment unit, not a burning elevator car.

In the face of uncertainty, one way to address this potential deadly circumstance is for firefighters to proceed on the basis that the fire is coming from an elevator car, not an apartment unit. Use a forcible entry tool to probe along the floor ahead of your movement. "Probe as you go!"

Unfortunately, a burning elevator car has resulted in the deaths of firefighters valiantly working to save lives and extinguish the fire. Don't forget that with an exposed hoistway you are just one step away from a deadly fall.

Now that we have discussed this very important safety matter, let's get back to other important points. Simultaneous with firefighting operations, other firefighters should check and remove passengers from other elevator cars threatened by the ensuing fire. This may not be a matter of concern if these elevators already have been recalled to the lobby by actuation of a smoke detector or Phase I recall of the elevators.

If the building is protected by an automatic fire sprinkler system, the actuation of sprinkler heads may already have "knocked-down" or controlled the fire. However, there is one concern regarding the sprinkler system that firefighters must be aware of. That is, if there is a shunt trip installed to kill the power to elevators in the same hoistway as the burning elevator car on actuation of a heat detector in the machine room or hoistway, power to all elevators serviced from that machine room is removed. Unless the elevators have already been recalled to the designated floor, firefighters could become dangerously trapped inside the cars. Another concern about the burning elevator car that has moved to the bottom of the hoistway is fire extinguishment. At this point, there can be fire burning on an upper floor and also in the pit or perhaps somewhere in between if the elevator safeties actuated. Under these dangerous conditions, with the possibility of falling debris, ropes or other elevator equipment, firefighters must not enter the hoistway to extinguish the fire. It should be done from an elevator landing. Firefighters operating on the upper and lower fire floors must coordinate their activities to minimize risk of harm and to eliminate confusion.

Just because the elevator car has run down or otherwise moved down, it doesn't mean that all of the ropes and or counterweights have dropped as well. However, if all hoist ropes are severed, as well as the governor rope, the elevator safeties will *not* set. Therefore, the elevator can fall. Staying out of and away from the hoistway entrance and pit is the message to heed here. It's not uncommon for firefighters to get injured or killed at the scene of a fire after the fire is declared under control. Things can and do happen that can compromise one's safety. Remember "Murphy's Law." Ole Murphy likes to hang around the fire scene.

As a rule, do not enter any elevator car damaged by fire—this means during fire-fighting operations or overhaul. Stay out of the car! Firefighters can cautiously use pike poles or similar tool to expose hidden pockets of smoldering fire, while operating them at a safe distance from the car and hoistway.

Machine Room

Typically, a fire in the machine room will actuate the smoke detector that results in the automatic return of all elevators in that hoistway to the designated level. Firefighters should be able to determine that the cars have returned as they enter the elevator lobby. If any car has not returned, then firefighters should locate the elevator to ensure the car does not have any passengers.

Firefighters arriving in response to a report of a fire in an elevator machine room are likely to find an overheated motor, shorted electrical equipment in the controller panel, or burning combustibles illegally stored in the room. Unless firefighters have the key to the machine room, they should be ready to forcibly open the door.

Given existing conditions, firefighters may have to extinguish the trash fire from the machine room doorway before removing power from the affected elevator(s). Firefighters need to be especially careful of

rotating equipment unless and until they have killed the power. For the other types of fires, firefighters should be able to shut off the power before entering the machine room. Also, before entering the machine room, firefighters may decide to allow smoke to dissipate or use fans to facilitate needed ventilation in the interest of visibility and safety.

Because of the limited space inside the machine room, the safety officer should limit the number of firefighters who enter and engage in fire-extinguishment activities. Once the power is removed from the elevator(s), firefighters should manage the fire as they would for a similar situation involving either an overheated motor or electrical short or failure. Do *not* attempt to access the controller panel until after the main power to that controller is turned off (figs. 20–3 and 20–4). The mainline power switch is the same one used to remove power to an elevator and its associated equipment, including the electric traction machine and its controller.

Fig. 20–3. Controller cabinet

Fig. 20–4. Controller equipment

Fig. 20–5. The pit can be like a big trash can for discarded newspapers and other combustible items.

As with other elevator emergencies, after the incident is over, the power should remain off with lockout/tagout still in place and the machine room door reclosed and checked for proper locking.

Hoistway Pit

Over time, the bottom of a hoistway can become a collection point for combustible materials such as discarded newspapers. This can present a fire hazard if the pit area is not cleaned on a regular basis (fig. 20–5).

The main source of ignition used to be a discarded cigarette. However, today, with the prohibition of using smoking materials in buildings, the likelihood of having fires in the pit is less than in years past. When fires do occur, they are mostly minor in nature and tend to be more of a smoke nuisance than an imminent life hazard. However, firefighters should not take a pit fire lightly. The hoistway has the potential to act as a chimney and spread smoke and toxic gases quickly to other floors.

If an intentional fire is set in the pit using an accelerant, it can quickly become a major fire event. The immediate concerns to firefighters relative to management of fire include:

- Being aware of the possibility of falling elevator equipment

- Extinguishing the fire at a safe distance from the hoistway entrance

- Ensuring that there are no passengers on the affected elevators

- Removing power from the affected elevators

Mounting an aggressive, yet cautious, fire attack is likely to have the most impact on minimizing the potential consequences of the fire. Other fire department activities will usually require almost simultaneous execution in the interest of the safety of firefighters and building occupants.

Summary

Elevator fires are mostly overheated motors or other electric devices. Other times, firefighters may respond to find smoldering paper at the bottom of the hoistway. In relatively rare cases, fire involves the elevator car itself.

As with any elevator incident, having standard operating guides (SOGs) are needed to identify objectives and tasks, address safety concerns, and identify duties and responsibilities. Adhering to safety precautions at ordinary elevator incidents helps to ensure adherence during more serious elevator incidents such as a serious fire in an elevator car.

When the circumstances surrounding a fire are such that other elevators are affected besides the fire-related elevator, then firefighters must ensure that the elevators are checked for passengers. This is especially important if a serious fire develops with the potential for life-threatening consequences.

Avoid entering an elevator car that is visibly damaged by a serious fire. It might be so damaged that the car may suddenly run or move down, leaving the hoistway exposed. By following the safety principles and consulting with the elevator mechanic, the incident commander should be able to maintain a safe work environment.

Fighting a fire involving elevator equipment can be an uneventful, ordinary event. Yet, that once-in-a-lifetime elevator fire that causes an elevator car to run or move, exposing the hoistway, can be an unforgettable nightmare. Under this situation, firefighters must not only be concerned about their safety but also the safety of passengers who may be stranded and endangered in other elevators sharing the same hoistway. Have a plan and be ready to use it!

Review Questions

1. What is the major hazard that firefighters must be aware of when managing a fire that occurs in the machine room?

2. What dangerous situation could a raging fire inside an elevator car pose to firefighters?

3. What should firefighters do before attempting to overhaul a burned-out elevator car?

Field Exercise

Develop a standard operating guideline on managing a serious fire involving an elevator car.

Chapter 21
Case Studies
and Lessons Learned

In this chapter, we focus on three very significant fires that included the use of elevators by firefighters. Each of these fires had a great impact on the fire service across the country, and in particular, on their own fire departments.

The Willoughby Towers Fire, Chicago

The Willoughby Towers Fire
8 South Michigan Avenue
Chicago, Illinois
September 24, 1981

Scenario

A box was struck by the Chicago Fire Department for a "report of a fire" on floor 25. The first due units captured the system and took car #4 to the 23rd floor (two floors below the reported fire floor). On arrival, they left a firefighter in car #4, and entered the stairwell to walk up to 25. When they left 23, they either left a member or blocked the door open. The firefighters tried the stairwell door at 25 and found that it was locked.

1. The door was cool to the touch.

2. They heard no noises from fire floor.

3. They elected to go back down to floor 23 and to take the elevator to floor 25.

Hindsight is 20/20

What the firefighters (six members) didn't know was that car #1 had been burning fiercely. (It had been loaded with trash bags.) The car had burned so badly that its front hoist ropes failed, causing the car to sit in the hoistway at an abnormal angle. The rear hoist ropes started to elongate from the heat. The car slowly started to move downward in the hoistway, but not fast enough to trip the car safety device meant to stop the car in an emergency. Due to the distortion caused by the fire, the car finally ground itself into the walls of the ninth floor. It had banged and scraped its way down 17 floors. What awaited the firefighters at floor 25 was this:

1. A superheated lobby (1,000°F)

2. Total obstruction due to smoke—zero visibility

3. The open floor doors of car #1—directly opposite car #4

The stage was now set for disaster.

As the firefighters in car #4 move up to floor 25, the safety circuit on their floor door interlock was destroyed by the heat of the fire in the elevator lobby. Their car was now dead in the hoistway—nearly at floor 25.

The officer in charge (OIC) elected to wait for five minutes to see if help would come, but none did. No one knew they were in trouble because they did not have a portable radio. Unfortunately, this was still fairly common across the country at that time. Further conversations with retired members of the Chicago Fire Department from that era revealed that even at the district fire chief level, fire officers did not have portable radios. (A hand light thrown through an upper story window drew attention to the plight of the firefighters.—Authors' note). The OIC elected to take them out. They had full protective gear on, but they still had to face a 1,000°F lobby. As they forced the doors open, they crawled out to their left, following the instructions of the OIC, who was familiar with the layout of the building. After finding a stairwell, the OIC's head count revealed five men instead of six. They all elected to go back onto the fire floor to find the missing firefighter. He had crawled out to his left, but swung just a little too wide and crawled right through *the open floor doors* (left open when the damaged car #1 pulled away) and he fell 17 stories, to the top of the derelict elevator, to his death.

On re-entry to floor 25, another firefighter crawled through the same *open floor door* and fell to his death, to the top of car #1. Then a third firefighter walked into the darkened lobby and stepped into the open hoistway. By instinct he threw his arms out when he felt nothing with his feet, and he grabbed loose hardware and was able to pull himself back up to the landing.

The results were two firefighters killed in falls, four firefighters were burned. Their gloves burned off, and the knees of their pulled-up rubber boots burned from crawling on the floor. A number of firefighters were also injured in a structural collapse at the ninth floor, where the rescuers forcing cement walls to gain access to the hoistway were struck by falling debris. They thought they would be rescuing survivors in the derelict car at the ninth floor.

Could this disaster have been prevented? As we both were working on the job at the time in our own communities and actively involved in firefighter safety, we individually examined their actions. The frightening thing was that both of us felt that we would have followed their actions at that time. We both felt that the OIC had used caution and followed procedure, but still ended up in a terrible situation. A locked stairwell door was the cause of their situation.

Lessons learned

If a door is locked, *force it*. Never take an elevator to a fire floor without it being verified as safe by other members on that fire floor.

(Reprinted with permission from articles compiled from the Chicago *Sun-Tribune* and *UPI*-Chicago and the Massachusetts Firefighting Academy-Department of Fire Services Student handout "3502 and you!")

Following is the Illinois Fireman's Rule. It was a court case brought about due to the incident.

Illinois Fireman's Rule

McShane vs. Chicago Investment Corporation, 601 N.E. 2d. 1238 (Ill. App. 1 Dist. 1992)

Several Chicago firefighters responded to a fire alarm at a 38 floor high-rise building in the city. In attempting to locate the source of the fire, they entered a freight elevator and rode the elevator up to within a few floors of the fire. They then walked the remainder of the way where they encountered the fire on the 25th floor. The elevator operated without any apparent difficulty. After checking the floors above and below the fire, six other firefighters subsequently entered the elevator on the 24th floor and pushed the button to go to the 25th floor, the location of the fire. The elevator, at this time, however, did not operate properly and lurched to a stop. The doors would not open. The firefighters attempted to get the elevator to operate or the doors to open, but were unsuccessful. They also discovered that the emergency telephone in the elevator did not work, nor did the alarm bell ring. They used their equipment to try to break through the access hatch at the top of the elevator, but found that it could not be forced open, and the firefighters were hit with a blast of intense heat and smoke. By this time their air packs were nearly empty, and the men were forced to breathe the hot, smoky air.

Ultimately, four firefighters escaped, but two died in the event. The surviving firefighters and the estates of the deceased firefighters brought suit against the owners of the building. A jury rendered substantial monetary verdicts to the firefighters, but the judgments were reduced by half based on alleged contributory negligence on the part of the firefighters. Owners of the building appeal.

Held: The owners argue that the Fireman's Rule precludes the firefighters from any recovery and that the firefighter's own negligence was the superseding proximate cause of their injuries. The Fireman's Rule is a doctrine which limits the extent to which firefighters or other public officers may be allowed to recover for injuries incurred when, in an emergency, they enter onto private property in discharge of their duty. Historically, a firefighter was considered a "licensee" to whom a land owner owed no duty except to refrain from inflicting willful or wanton injury. Over time, however, courts began to recognize many exceptions to this harsh rule, finding that various circumstances raise the status of the injured firefighter to that of an "invitee." In Illinois, a firefighter has no right of recovery by causes related to the fire. He does not assume, however, the risk of being injured by causes unrelated to the fire, i.e., causes that might be faced by an ordinary citizen entering upon the property. A high-rise fire is somewhat unique because elevators must be utilized by firefighters to reach the fire scene with their equipment, despite the fact that elevators are inherently dangerous during a fire, and ordinary citizens are usually directed to refrain from using elevators during a fire. Consequently, an elevator is a place where a firefighter might reasonably be expected to be while fighting a fire in a high rise. Here, there is no question that the elevator failed to operate properly. A review of the evidence in the case reveals that the conditions under which the elevator and its communication and alarm systems were constructed were unsafe, and that the defects form a proper basis for finding the defendant negligent. The conditions were not caused by the fire and are, therefore, not fire related so as to be covered under the Fireman's Rule. The building owner argues, however, that any negligence attributed to it should be negated by contributory negligence of the firefighters in actually taking the elevator to the 25th floor. Firefighters must utilize elevators to reach the fire scene in high-rise situations, and the use of such elevators should not be indicative of a firefighter's negligence. Nevertheless, it is a cardinal rule of the firefighters, taught to them during their training, to never take an elevator to the fire floor. In the present case the jury apparently found that the firefighters' own negligence in using the elevator made them contributory negligent sufficient to reduce their recovery by 50%. The allocation of the various parties' negligence is a matter within the province of the jury and should not be disturbed by the court. Consequently there is no reason to upset the jury's apportionment of liability between the parties. Affirmed for firefighters.

Fire Service Labor Monthly (January 1993) vol. 7, no. 1.

Fireman Fighting Arson Survives 14-Story Fall

By Mike M. Ahlers
Journal Staff Writer
Montgomery Journal, October 29, 1981

A 24-year old volunteer fireman who survived a 14-story plunge down an elevator shaft while fighting a Rockville fire early yesterday remained in critical but stable condition at the Shock Trauma Center in Baltimore yesterday afternoon.

David Flower of the Kensington Volunteer Fire Department in Glenmont walked into the open shaft while fighting the two-alarm smoky blaze which routed several hundred residents out of the 17-story Rock Creek Terrace apartments on Veirs Mill Road.

Fire officials said the fire was arson.

A total of 24 persons were injured, two of them seriously. The second person seriously injured was identified as Susan Nugent, 25, a resident of the apartment. She was listed in critical but stable condition at the Washington Hospital Center with second and third-degree burns over 11% of her body.

Fire officials say the fire apparently was set by use of a flammable liquid in the freight elevator at the southern end of the high-rise. When the elevator doors opened on the 12th floor, flames and smoke spewed from it, officials said.

At some time during the fire, officials said, the flames ruptured the elevator cable and it dropped to the subbasement about 14 floors below.

When firefighters arrived on the 12th floor, they discovered a woman apparently overcome by smoke lying on the hallway floor and carried her to safety.

Officials said Flower then took the lead end of the fire hose and advanced down the smoke-filled hallway.

"The elevator is at the end of the hall, so with the doors open and the smoke obscuring (Flowers' vision), he just walked to the end of the hall and fell," Kensington Chief Fred Bagley said.

"He either thought it (the elevator opening) was an apartment or it was a continuation of the hall," Bagley said. "We're still not sure why he didn't (die)."

Bagley said officials speculated that Flower either "ricocheted" off the walls of the shaft and was slowed down, or that his fire gear and debris on the top of the elevator cushioned the 12-story fall.

One of Flower's boots had come off indicating that his leg had become entangled in the elevator mechanism, Bagley said. A fire official said Flower narrowly missed being impaled on a rod on the elevator.

After fellow firefighters worked roughly 10 minutes to free Flower, and then loaded him into an ambulance, several of his distraught friends threw their fire helmets to the ground and fought back tears.

Flower was flown by helicopter to Baltimore where he was treated for fractured ribs, internal bleeding in the chest and stomach area, a broken jaw, and second-degree burns.

Six other persons were taken to Holy Cross Hospital for smoke inhalation and burns including Kensington firefighters James Abell and John Feissner, Bagley said.

Six police officers were treated at the scene for smoke inhalation they suffered while helping evacuate the 10th through the 17th floors. Officials estimated that as many as 70 people were evacuated from the building.

The Rock Creek Terrace complex has been the scene of numerous "nuisance fires" and false alarms, according to residents. One resident of the 13th floor, Dale Simpson, said he had trouble convincing people yesterday that the fire was real. Simpson said he banged on doors and warned occupants on the ninth floor, where his sister lives, but "nobody would believe me," he said.

Lessons learned

Once again, *an open floor door*, obscured by the smoke of a fire that he was fighting, nearly cost this young firefighter his life. He certainly was a very lucky man to have escaped this fall with the serious injuries he incurred, which could have been much worse.

When in smoke, *crawl!*

Dolphin Cove Condominium, Clearwater, Florida

255 Dolphin Point Road
Islet Estates
Clearwater, Florida
June 28, 2002

The scene of this early morning fire resulted in two civilian deaths and five firefighter injuries, some with serious burns. This fire was covered extensively in the USFA report #USFA-TR-148/June 2002[1], from which material for this chapter has been used. It should be read in its entirety to fully understand the fire conditions and the operations of the Clearwater Fire Department (CFD) at this incident.

As a lessons-learned opportunity, it has many features in common with the Willoughby Towers fire.

- A locked stairwell door played an important part in decisions to take an elevator by some of the firefighters.

- Firefighters took three elevators directly to the reported fire floor.

- Firefighters were temporarily trapped in one elevator due to the elevator floor door suffering distortion damage from the fire.

- All members wore full protective gear and still suffered burns and injuries.

- Members deviated from a CFD-written standard operating guide, which prohibited the use of elevators for reported fire floor below the sixth floor.

Summary

One of the problems that we, the fire service (fig. 21–1), continue to display is making the same mistakes over and over again. In the past few years, we have seen disastrous losses of firefighter lives in major fires (Storm King Mountain, Colorado, 1994). Unfortunately, it was a mirror image of a past forest fire that claimed firefighter lives in nearly the exact similar situation as one that happened several years earlier in Mann Gulch, Montana, in 1949.

Fig. 21–1. Clearwater, Florida fifth floor elevator lobby (photo by permission of USFA and TriData Corporation)

or smoke detector (FAID) that no one will replace, or the same water surge alarm because two people in a building flushed at the same time, and we end up with the biggest killer of firefighters—complacency. The awareness of danger is down, and they expect to be back in quarters (BIQ) shortly. Then the doors open onto a fire- and smoke-involved elevator lobby, and it is all over.

It is your job, and ours, to make sure we are not writing about a *fourth* mirror-image fire involving members, elevators, and fire floor landings.

Review Questions

1. Have you ever taken an elevator directly to a reported fire floor?

2. If so, how did you try to justify endangering the lives of your company members to them?

3. Did that building have Phase I and Phase II?

4. Does your department have an SOG for elevator operations during a fire?

Field Exercise

Locate a multiple-story building in your district. Visit that site and determine the situation relative to the following:

- Does it have firefighters' service?

- Determine the situation relative to the stairwells: number, location, and locked/open.

- Create a status page for your company response guide book, with the information you have gathered.

Endnotes

[1] "Multiple Fatality High-Rise condominium Fire." Clearwater, Florida. USFA-TR-148/June 2002.

We all attend classes that show us what happened, where and how their lives were lost, and we leave saying to ourselves, "I'll never make that mistake." Yet, we see it clearly in results of the three fires that we are examining in this chapter. Every new kid on this job coming out of the local training academy or city recruit program knows the following:

- We never make a direct approach to a reported fire floor in an elevator.

- We always go two floors below the reported fire floor.

- Depending on local preference, we walk up the stairs for sixth floor or lower reported fire floor, after securing the elevators on Phase I.

- When in smoke, crawl!

Yet, we all know what happens in the field. Companies go to the same building for the 33rd time in a month, for the same fire alarm initiating device,

Chapter 22
Shunt Trip Activation
and Its Threat to Firefighter Safety

In jurisdictions not enforcing the NBCC, means shall be provided to automatically disconnect the mainline power supply to the affected elevator upon or prior to the application of water from sprinklers located in the machine room or in the hoistway more than 600 mm (24 in.) above the pit floor. This means shall be independent of the elevator control and shall not be self-resetting. The activation of sprinklers outside of the hoistway or machine room shall not disconnect the main power supply line.

This paragraph is from the A17.1–2004, section 2.8.2, Safety Code for Elevators and Escalators. To the average firefighter, it is just another dry code section that covers, among other things, sprinklers in the elevator machine room. To be unfamiliar with its potential consequences could be very dangerous for those firefighters who ride Phase II elevators up into buildings to fight a fire.

The presence of sprinklers requires a means to meet the requirements of A17.1, section 2.8.2.3.2. This action described is referred to as *shunt trip*. A shunt trip is an electrically operated circuit breaker (fig. 22–1). The installation of a shunt trip breaker allows the power to be shut down on a signal from another source, such as a heat detector. (NFPA 72, 2002 Edition, section 6.15.4.1).

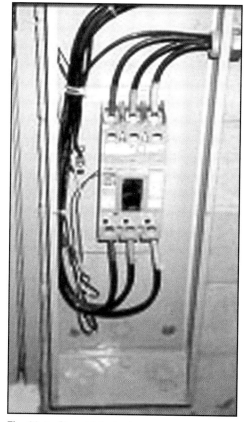

Fig. 22–1. Shunt trip device

The important codes and standards and the referenced paragraphs that impact these installations are:

- A17.1–2004, Safety Code for Elevators and Escalators, section 2.8.2.3.2 (which requires removal of main-line power from the elevators)

- NFPA 13, 2002 edition, *Standard for the Installation of Sprinkler Systems*—the recognized expertise when it comes to sprinklers in buildings

- NFPA 72, 2002 edition, National Fire alarm Code, section 6.15.4—dictates how elevator shutdown is to be accomplished

- NFPA 70 National Electrical Code, 2005 edition, article 620–51—additional requirements for disconnecting power to elevators

It is critical to know which version or edition that your local authority having jurisdiction (AHJ) has adopted. It is even more important to be familiar with the contents and how they impact your buildings.

For many years, automatic sprinklers have been saving lives and property in buildings across this continent. It has been a long, hard fight to convince the powers that be that sprinklers would save them money in the long run. This savings would come from lower insurance costs and construction trade-offs in other areas. We, in the fire service, were among those whose voices were not listened to, as buildings, now known as high-rise or tall buildings, were thrown up everywhere without automatic sprinkler protection. What was foremost in the minds of the developers was how much money they could save by cutting out *needless* (sprinklers!) expenses. Ironically, it has been some of those same unprotected buildings that have been the scenes of significant high-rise fires, for example, the First Interstate Bank Building, Los Angeles, California, on May 4, 1988, when ignition occurred during the retrofit installation of sprinklers that had been mandated by AHJ.

As automatic fire sprinkler protection became more the norm in the erection of buildings of all sizes, the elevator machine room was usually not sprinklered. The machine room is a site of very high electrical activity, as well as moving machinery that may not operate safely

if water from a sprinkler head discharged on it. Some of the dangers presented to the life and safety of the passengers include:

- A wet brake being unable to stop or hold a car from running *up* into the overhead, causing injuries to occupants

- A wet brake being unable to stop or hold a car from running *down* into the elevator pit and buffers, causing injuries to occupants

- The water spraying onto sensitive electrical components, creating bridging, or short circuits, which, in turn, could make the elevator run erratically and uncontrollably

- Premature shutdowns, trapping occupants or firefighters

For many years, this exception to their installation was found in most elevator machine rooms. The practice now is to have all spaces in a building fully covered by sprinkler protection. The fire service is now left in a quandary, in that we are always in favor of any means to protect the lives of the public, including in this case automatic fire sprinkler protection. Unfortunately, in this application, the sprinkler protection presents a considerable obstacle to the safety of the firefighters.

Predetermined marooning—The conscious decision to immediately stop cars where they are when shunt trip operates, whether they are at a landing or between floors. It was believed by those concerned that firefighters could easily force their way out, which is not the case.

When the decisions were made about shunt trip, all the vested interests voiced their (NFPA 13, ASME A17.1) concerns, except the fire service. As mentioned numerous times in this book, we, the firefighters who will be using these elevators, have to be at the table to have our concerns heard and acted upon. The installation of sprinklers in the machine rooms satisfied the NFPA 13 interests, in that any fire condition could be controlled or extinguished. The elevator industry was very concerned to see that water was going to be potentially flowing in their machine rooms, so shunt trip operation was developed and installed to protect the riders and equipment from the hazards mentioned earlier in this chapter.

At this point, we have to ask, "Where does this leave us, the firefighters, if the shunt trip actuates while we are riding the elevator during Phase II?"

- It leaves us stuck in the elevator, between floors more than likely.

- It leaves us facing the additional challenge of overcoming the restrictor.

- It leaves us trying a forcible exit from an elevator that is stalled in a hoistway!

- Might we add, it leaves us in a building that is also on *fire!*

As firefighters, we accept risks involved in entering a burning building. That is all part of our profession, but we do it with a knowledge that we are all that there is between the fire and the people. There is no one else going to do this job, and parties have elected to maroon us, between floors, with the hope that we can perform *forcible exit* from the car. At all times we should try to use the normal car exit door to escape the entrapment. If forcible exit is necessary, using the car top exit is our next resort, although not recommended, due to the fall hazard that exists out in the hoistway.

Some jurisdictions have elected to modify the shunt trip system to make it safer for their firefighters. Listed here are examples. Yet, the important issue remains that you must become familiar with what your local AHJ has adopted or modified for your jurisdiction. Remember, not everyone has adopted the latest editions of the ASME A17.1 Code, and most others have never adopted ASME A17.3 Safety Code for Existing Elevators and Escalators.

Example 1:
Seattle Fire Department

Administrative Rule 9.08.05: High-Rise Buildings

2.1.2 A *solenoid* valve shall be installed on the sprinkler supply line, and shall be in an accessible location *outside* the machine room. The solenoid valve shall be of the normally open type and shall be energized under normal conditions to prevent water from entering the machine room piping.

Other Than High-Rise Buildings

2.2.1 An approved, *manually operated* valve with an integral switch shall be installed on the sprinkler supply line for each elevator machine room. The switch shall be connected to the elevator power disconnect device. The valve shall be easily accessible and located *outside* of and next to the machine room door not higher than six feet above the floor. The valve shall be normally closed. Opening the valve shall shut off power to the elevators and charge the sprinkler lines with water.

Note: The previous two paragraphs are excerpts from a major document. Read the entire Rule 9.08.05 to understand its full impact.

Example 2:
Commonwealth of Massachusetts

Excerpt from 524 CMR 17.02, Sprinklers and Sprinkler Piping

Effective January 1, 2004, section *524 CMR 17.02(20)(f)(1)* required that all sprinkler heads and the associated piping must be removed from all existing elevator machine rooms, hoistways and pits. The sprinkler piping must be cut and capped outside the wall of the machine room, hoistway and/or pit and removed.

Shunt trip device

Effective January 1, 2004, section *524 CMR 17.02(20)(f)(2)* required that the automatic mainline power disconnecting devices (shunt trip device) shall be disabled, disconnected, or removed.

When a State Elevator Inspector finds sprinklers and sprinkler piping in an elevator machine room, hoistway and/or pit or a shunt trip device that is not disabled or removed, the inspector typically gives the building owner 180 days to come into compliance.

Note: These two paragraphs are excerpts from a major document. Read the entire Code of Massachusetts Regulation-CMR 17.02 to understand its full impact. On September 9, 2005, this change became permanent in 780 CMR, section 904.1, Exception # 3.

One might say that both of these examples are drastic steps. Firefighters believe that automatic fire sprinkler protection is like apple pie and motherhood—subject areas never to be abused. But we also value the lives of other firefighters and do not require dead firefighter bodies to justify our fears about entrapment in an elevator via *predetermined marooning.*

Firefighters who have buildings with shunt trip must decide now how they would manage escape from an elevator stalled by the actuation of shunt trip during a fire. Leaving the decision to desperation should they become trapped, firefighters might not have the time nor the wherewithal to successfully escape.

Summary

The actuation of shunt trip is not a new invention, but one that has been around for many years. The problem is that most firefighters are totally unaware of it being in place in the buildings of their district. The latest *proposed* changes to the ASME A17.1/CSAB44, Elevator and Escalator Safety Code, and NFPA 72, National Fire Alarm Code, have made provisions to have timers installed to hopefully allow the completion of Phase I recall before the shunt trip drops the axe and maroons the car. At the time of this writing, this is not an accepted rule yet, so you will have to consult with your local AHJ to find out if this change has become part of the standards.

It should not be lost on the members of the fire service that all the fire service requested was to have the car stopped at the next available floor and to be let out before shunt trip operated while we are on Phase II. Attempts were made to request consideration for the elevator machine room to be considered under the NFPA 13 exception for electrical rooms. This exception, if granted, would have eliminated all of the problems of shunt trip! It was denied on more than one occasion.

There are thousands of shunt trip installations across the United States, all put in with good intentions, but which will create an undue hardship on firefighters already in a perilous conditions. It is important that you find out which of your buildings has these units and plan accordingly.

Review Questions

1. Define the shunt trip system and how it operates.

2. Relate the Phase II problems that can be anticipated because of it.

3. What is another name for NFPA 70?

4. Describe predetermined marooning.

Field Exercise

During company in-service inspections, locate two elevator systems with a shunt trip in your first-alarm response district and preplan your escape. Request assistance from the responsible elevator company and the building management with this exercise.

Chapter 23

The Firefighters' Emergency Operation Key: FEO-K1

The time has arrived for the common key. A very welcome proposal has been submitted by the National Elevator Industry, Inc. (NEII) to the ASME A17.1 Emergency Operations Committee. It requires a common designated key in the ASME A17.1 Safety Code for Elevators and Escalators and CSA B44 Safety Code for Elevators this year (2006). This proposal was successful, and now requires a common key for Firefighters' Emergency Operation. This key will be known as the FEO-K1, and will be a tubular 7-pin style 137 construction key. It was passed at the Standards Committee meeting of ASME A17.1, and the proposal would allow for continued use of existing common key systems, such as 3502 and others already in place. This advancement is a direct result of the input from firefighters and elevator industry members at the "Workshop in Atlanta," which was held in 2004. The findings from that Workshop have been forwarded to the respective code committees to bring about the needed changes.

As members of the fire service for many years, we have seen the problems associated with the operation of Phase I and Phase II Firefighters' Emergency Operation first hand. A major roadblock to the effective operation of this very worthwhile system has been the numerous keys most fire companies have to contend with upon arrival at various buildings within their own cities or jurisdictions. That is, *if* there are keys available to use, or if management can find them, and so forth. For many years there has been a resistance to the concept of a common key. When you spoke about the concept of one key, you heard horror stories of potential abuse of the key.

The National Fire Protection Association (NFPA), International Fire Service Training Association (IFSTA), and others look to the ASME A17.1 Safety Code for Elevators and Escalators and CSA B44 Safety Code for Elevators. It is important to remember that most of the organizations that publish fire related manuals and books are just that—*publishers*, and not firefighters. There may be retired or contract people involved who are firefighters, but they seldom if ever decide policy.

The result of this has been that many fire companies have no faith in the system. The common complaint across the kitchen table in firehouses is:

- They never can find the key upon arrival.

- The key that was "guaranteed" to be at the desk is not there.

- It is the wrong key.

- The cars and the system were modernized, and the new manufacturer did not leave keys.

- There is *one* key, and there are *eight* elevators.

- The key does not work in the switch.

Good pre-fire planning, and getting out into your districts, asking to see the keys and trying them in the tumblers to see if they work properly, can lessen the effect of these conditions. Waiting until the night of the fire is too late. Place the keys in the security box on the building where your community places all of the other secure items from a building.

A common firefighter service key that the firefighter can carry has the following benefits to *all*:

- Immediate access to the elevators by the fire service. No trips to find "Ernie the custodian" or the security chief.

- All firefighters in a jurisdiction (city, county, or state) can operate in other jurisdictions' buildings when operating in either move-up or mutual aid situations in neighboring districts or communities.

- The keys can be duplicated and given to *all* firefighters in the jurisdiction to keep with their individual firefighting gear.

- This fulfills a Department of Homeland Security requirement (Homeland Security Presidential Directive #5) of *interoperability* during mutual aid, which is a major demand of the federal government in anti-terrorist preparations and funding.

A major hurdle to overcome has been from within our own ranks and those who enforce fire prevention codes. There is a misconception that there will be abuse of the privilege of having the key. Fears of unauthorized personnel taking elevators, causing disruptions in service, is just one of the reasons given by opponents to not have the common key. We would point out this fact to our firefighter brothers and sisters: Gamewell Company is a well known manufacturer of fire alarm boxes that we see on street corners and buildings with an interior fire alarm system. Unless they are asked to key the lock differently, they ship these fire alarm boxes all over the world with the *same* "Christmas tree" key.

There have been cities and states that have long had a common key for fire department use, such as New York City with its 1620 key. In Massachusetts and most of New England there has been a common key, the 3502 key, since the 1980s. Fire apparatus in Provincetown, Massachusetts on Cape Cod can travel nearly 300 miles to Pittsfield, MA, and still use the 3502 key from their apparatus to operate in the buildings of that distant community. Presently, the City of Orlando, Florida has adopted the 3502 key for itself and several other major jurisdictions in Florida. The City of Toronto and the City of Atlanta and the State of New Jersey are also examining the question of a common key. In Massachusetts, only firefighters, elevator mechanics, and state elevator inspectors may possess the 3502 key by law.

During the time that the key has been in place:

- There had been cases of abuse by security guards and ambulance personnel who received the keys from elevator people and firefighters. These instances are now few and far between.

- By enforcing the penalties, that problem is now at a minimum.

- When people complain about abuses, ask them for the *documentation* of the incident. Documentation provides a means to correct or disprove that a problem is present. Stories of abuse are like firehouse stories about fires, in that they get bigger and worse each time they are repeated, but all should be investigated.

Some of the important gains for *everyone* concerned by using a common firefighters' key are:

- Firefighters will actually use the system, rather than walk by it because of the problems encountered in the past.

- Mutual Aid assignments are able to use the elevators

- The daily use of the system while on Fire/EMS calls into buildings enhances their knowledge and faith in the system.

- The firefighters will become more knowledgeable as they use the one-key system.

- A very good system developed by the hard work of many elevator people will finally be used to its full potential.

By having frequent usage, problems can de detected and reported sooner.

When a feature is used frequently, it performs better because of the regular use of its components. A system that is there for firefighters to use will actually be used in a productive way.

In the future, unless your AHJ has modified their code, the supplied key with the elevator components will be the FEO-K1. This is a great step forward.

Fig. 23–1. The common 3502 key

Summary

The Fire Emergency Operation Key (FEO-K1) will now be part of the delivered system in all new elevators installed under ASME A17.1 Safety Code for Elevators and Escalators-2006. This is a major development that will resolve many of the complaints and problems that have been associated with Firefighters Emergency Operation. As with all complaints, the solutions involve all parties concerned, and that includes the members of the fire service. We must get out of the stations and into our buildings and make sure that all of the components work. This means before the fire, and to provide you with a means of familiarization with another tool of the job. The bureaus of Fire Prevention, Fire Suppression, and Fire Training must work together on this project as they do, or should, on all matters pertaining to the member's safety on the fire ground.

Review Questions

1. What is the FEO-K1?

2. Which Phase does it operate?

3. Where did the concept develop from?

4. What is the relative code?

5. Who is responsible to get the message out?

Field Exercise

Working with the local elevator AHJ, visit and familiarize your self with a new FEO-K1 site and an older non-common key site. Start making up note pages for your response book for all groups in the company to use.

Chapter 24

Use of Elevators During a Building Fire

Fig. 24–1. Clearwater, Florida, fifth-floor elevator lobby, 2002 (photo courtesy of USFA[1] and TriData Corporation)

One of the most dangerous operations the fire service performs is the use of elevators during a fire in a building. Using an elevator is the *third* most common means of transport used by firefighters, preceded by fire apparatus and stairs. Our history is full of incidents where for one reason or another, firefighters have been killed, seriously burned, or injured while using elevators during a fire. Because there are many versions of the ASME A17.1 Code Firefighters' Service in existence across the continent, firefighters are faced with a number of variations that impact their trips to the upper floors. A fire company will find different generations of firefighters' service in numerous buildings, even within their own response district. The questions they must ask of themselves include:

- What version of firefighters' service is installed in this building?

- Does this building even *have* firefighters' service?

- Do we know how to operate the version we have?

The obvious is usually in front of us, with smoke or fire showing from an upper floor and reports of people trapped on or above the fire floor. Some very important aspects must be covered at this point about the familiarity the members responding may or may not have with the particular building. It is expected by the rest of the response that the fire companies who are stationed in the area will be familiar with the building. Tragically, it has nearly always been the first due engine company and ladder or truck company that has become trapped either in the car or on the fire floor landing. Sometimes this happens to the best-trained fire companies when equipment fails, but at other times it is because members fail to follow established procedures. The rest of the first-alarm fire companies seem to make their way up to the fire area without a problem, because they either know from radio transmissions what the conditions are upstairs, or have been warned by the situation unfolding. They also could be using elevators in a different bank or group of elevators, or simply using the stairs.

Horror stories are told at the table in firehouse kitchens about near or actual disasters involving fire company members responding to the upper floors in elevators. Occasionally, a fire department will let the true story out to the rest of this "band of brothers" to learn from their department's tragedy. The lessons learned from the tragic results of the fire at The Regis Tower Fire, 750 Adams Street, Memphis, Tennessee (4/11/94), which were published for all in the fire service to learn from, is an example of leadership at the top that we all can look up to (Inquiry Review Board ordered by Charles E. Smith, Director of Fire Services).

Unfortunately, as we all know mistakes and errors often get buried in the paperwork, never to see the light of day. The cardinal rule of the firehouse is:

- Keep it in the group (shift).

- Keep it in the station.

- Keep it in the battalion or division.

- Above all, *never* let group 2 know about it!

The lessons learned that are so vital to our learning process are lost to all who could benefit from them.

Upon arrival at the incident, the company officer arriving "first due" should be checking the fire alarm initiating device (FAID) annunciator panel for the floor of activation. We should never take the word of building security, but go check the panel ourselves. It may be located in the lobby or in a fire command post, which may be conveniently located at the entrance, or as much as a 1/4 mile away, tucked away in another part of the building. As we showed in chapter 5, Hallway and Elevator Lobby Features, the lobby contains many elements that can assist us in determining what the conditions are in the building above.

After you know where you are going, who is going with you, and what your assignment is, determine if you have sufficient firefighters and equipment to perform the job the incident commander (IC) has just given you. Do not be afraid to tell him that you need more help or equipment to perform this task he has just handed you. The successful extinguishment of a fire depends on many things, but if any one of those tasks is not completed or not feasible, then the whole extinguishing effort suffers.

As the companies assigned to make the first movement up the hoistway in the elevators, members must be fully equipped with full firefighting gear, including self-contained breathing apparatus (SCBA) and portable radios. Another term used in describing firefighting gear is personal protective equipment (PPE), meaning a full set of firefighting gear! This equipment should have been put on before stepping off the apparatus on arrival. The requirement of entering a fire building and using elevators for transport to the upper floors has its own set of rules that apply to the necessary tools and other items of equipment we have to carry. All of these decisions should have been made via standard operating guides (SOGs), developed and enforced prior to the fire. Most SOGs, as they are called, have been written based on prior experience gained with other fires and problems that the department wanted to correct before the problem or crisis recurs. Many departments write their own to fit the local situation, gleaning the basic components from model SOGs that are available to all on the Internet.

Most communities do not have the staffing that we might see in major cities such as New York, Los Angeles, Chicago, Atlanta, or Boston. The fire does not care what your staffing limitations are, and it will not show any consideration for your local problems. It will kill and injure your members unless you build and *enforce* SOGs to protect them and work within your own staffing limitations.

Fig. 24–2. Elevator lobby

The following questions should be asked:

- As you enter the elevator lobby with your fully trained and equipped crew, have you determined, *What is the status of the elevators?*

- Does this building have a full firefighters' service operation, Phase I and Phase II in its elevators?

- Are they all working and available today?

If the fire is reported to be on the sixth floor or lower, do not use the elevators to attack the fire. Use the stairs *after* you have either confirmed that all of your cars are down on Phase I automatic recall or you have captured them with the Phase I key and they are emptied of passengers.

Since the first installation of firefighters' service in 1972, many improvements and changes have taken place in the design and function of this system. The initial installations were at 70 feet or higher, which was then changed in 1989 to those over 25 feet. Later this was changed to 2,000 millimeters or 80 inches, and the hoistway does not penetrate a floor. Please keep in mind that the ASME17.1 Code, unless a retrofit has been mandated by a legislative change, does not create updates to the new code for the existing elevators.

To borrow a term, this is not your grandfather's— or even your father's—elevator. The average life span of a modern passenger traction elevator is about 20 to 30 years. At that point, the elevator may be replaced with a new car during a modernization project, and new controllers and other parts are installed. With that modernization, the improvements that have been developed and are now part of the code will be in place in the newer equipment. The elevator ropes (wire steel) are usually replaced every 10 years or earlier if needed because of workload. The elevators and other built-in fire protection and detection systems in today's buildings are a far cry from what we faced in the 1970s and 1980s. During that era, it was not *if* you would lose the elevators, it was *when* you would lose them! The design of the systems at that time allowed for smoke and heat to easily move from the fire floor up the hoistway and into the machine room. The elevator hoistway is the best chimney in a building, and the openings in the machine room floor (see chapter 3) allowed the products of combustion to attack the controllers and machine components. Countless reports from that time revealed that the elevators were either inoperable or operating erratically and not usable on arrival of the fire service. Please keep in mind that there are still plenty of the older buildings and elevators out there in the field. They have not gone away, but they have just gotten older!

Fig. 24–3. Lobby FAID

The most important change from a life safety viewpoint is the development of automatic recall of the elevators by the dedicated FAID in the elevator lobby at each landing (fig. 24–3). This feature alone has saved more lives than all of the other fixed systems in today's building. Before you have finished pulling up your bunker pants in the station, the elevators are down in the lobby on Phase I automatic recall operation. There have been many changes and improvements via design, material, and code changes, all learned from the fire behavior of the elevators in the past. All of these efforts and changes have helped make the elevator trip safer for us than it was in the past, but there are still no guarantees that the elevator will not fail during a fire. The fire environment is an enemy that can destroy everything that humans have devised, and it is unforgiving.

It is too late for you to begin determining what is inside this building and its elevators as you pull up to the curb. The secret to knowing what is going on outside of the firehouse window is to get off the apparatus when on a road trip for supplies and provisions, and go into the buildings for a preplanning session with the building engineers or maintenance staff. These are the people who can give you the answers to the questions that you will be asking yourself or anyone else who will listen at 0345 hours the morning of the fire. A modern building will have automatic recall on the elevators, so that if smoke or heat activate an FAID in the lobby of the group of elevators you are dealing with, they will all automatically return to the *designated level* (DL; the lobby) or the *alternate level* (determined by the authority having jurisdiction [AHJ]) on Phase I automatic recall (fig. 24–4).

Fig. 24–4. All cars *must* answer recall to designated landing (DL)

If the elevators have not recalled automatically, then you should do so manually at the lobby by operating the Phase I firefighters' service key switch. All cars must answer the recall, and if they do not, report this to the IC. Any elevator that does not recall must be considered as missing with the lives of those in it at great risk. A rescue group would have to be assigned the task of finding the elevator and removing those passengers in the elevator. Every firefighter should think of a missing elevator as an attic apartment, or the back bedroom in a tenement. If we had a report of people missing in those situations, we would fight our way to them. We must think of the missing elevator in the same vein.

It must be remembered that the use of elevators during a building fire plays a very important yet smaller role when placed against the entire firefighting operation. This chapter will take you from the curb to the landing on the floor two floors below your indicated fire floor. After that, it is into the stairwells and up to the fire floor door, checking it for heat, and make your entry.

Precautions to the Use of Elevators During a Fire

As this chapter is on the use of elevators in a fire building, and the precautions we must exercise, we must adhere to the following rules:

- After determining the floor of activation from the fire panel, move to the elevator lobby.
- Use the firefighters' service key to operate the Phase I recall if the elevators have not recalled automatically. Some departments carry a uniform standard key to facilitate this action. (New York: 1620 key; Massachusetts, most of New England, and Florida: 3502 key; others may use a local AHJ key) (fig. 24–5).
- Count all of your elevators to make sure they have all answered the Phase I recall. If any are unaccounted for, notify the IC immediately.
- Check hoistway for smoke or fire conditions by looking up between the car top and doorway opening. Shine a hand light up and look for accumulation of smoke in the hoistway.
- Check the Phase II station on the car operating panel (COP) in the elevator to make sure that the fire hat symbol is not flashing (fig. 24–6).

Fig. 24–5. Phase I lobby capture stations. New with reset, and older with by-pass (courtesy of C.J. Anderson Co.)

Fig. 24–6. Phase II key switch on car operating panel

- A flashing fire hat emblem means the FAID in the elevator machine room for that elevator bank has been activated by smoke or heat. At this point, *if the fire hat is flashing,* a decision must be made whether to use the cars in this bank or not. That decision is up to the IC.

- Remember that you may have a *shunt trip* situation coming!

- Depending on the local SOG, use elevators in a hoistway that do not pass through the reported fire floor. Some departments allow it with caution; others prohibit it.

- A blind elevator shaft is the only one that could be used by firefighters to pass by the fire floor to go above the fire. There are no floor door openings in the shaft at the lower levels during the blind hoistway run. This is per your SOG.

- Avoid the use of service or freight elevators until they are declared safe to use. This will be the decision of the IC, after he or she has evaluated the information. Freight elevators usually are located in the back or service areas and have, in many cases, proven to be located near the source of the fire.

- Follow the rule of six: Do not use the elevator if fire is on the *sixth* floor or lower, and no more than *six* firefighters are to be permitted in any elevator at the same time. Make sure there is still enough room to allow them to quickly don their SCBA should things suddenly "go to hell in a handbasket" when and if the fire tries to join the group!

- Following the rule of six will provide a quicker movement to the fire floor via the stairwell and prevent overloading of the elevator.

- Per the lobby control officer, have members with proper gear enter the elevator, and assign a member as the driver, or "taxi."

- There must be at least one portable radio and preferably one on all members in the car, or it cannot leave lobby control.

- SCBA should be maintained in the ready position.

- Allow no one except fully equipped fire department members in this elevator. This *does* mean "Ernie the custodian" *cannot* come up with you.

- Make sure your crew and the driver (taxi) are familiar with the components of this COP. Does everyone know what *call cancel* and the *flashing fire hat symbol* mean? Does everyone know how to set the emergency stop switch on a moment's notice,

- Depending on installation date and local AHJ requirements, is an emergency stop switch available to those in the car?

- Do you have the necessary forcible entry/exit tools and firefighting equipment with you?

    1. In the event the car does not stop at the intended floor, pull the door open. If necessary, a tool may be used to pry the elevator door open, thereby disengaging the car door contact switch.

    2. In the event that the car should become disabled, a set of *forcible exit* tools may be needed to allow self-extrication by the members. The term *forcible exit* is defined in the Glossary.

    3. Forcible entry tools may be required to gain access to a secured area after the car door has opened.

- While pressing the floor button for at least *two floors* below the reported fire floor, also press the door close button, keeping the pressure on until the doors have fully closed.

- A conservative approach is to go *five stories* at a time, visually checking above for smoke or fire, and listening for fire sounds from above. A new floor designation will have to be entered after each stop. Take a few seconds to check the hallway to locate the stairs, just in case it becomes necessary to escape from the fire floor.

- As your elevator slows to glide into your floor and stops, the doors will stay shut to protect the members. It will require the driver (taxi) to push the door open button with a momentary pressure, while another member checks the lobby that will be visible for an instant.

- Because the driver (taxi) only used *momentary pressure* on the button, the door opened an inch or two, then swiftly closed. To complete the door open sequence, press the door open button with continual pressure until it opens fully and stays open.

- If the lobby is smoke free, then the driver (taxi), at the company officer's order, can open the car door fully to allow the members to exit onto the floor.

- This is the *discharge floor*, and it is a safe floor for arrival of the other cars. They should all be on firefighters' service Phase II Operation throughout the fire incident.

Take the following precautions on elevator use:

1. Determine as soon as possible if the location of the fire could affect elevator operation.

2. Any indication of water, smoke, or fire conditions existing in the hoistway requires that another bank of elevators be used or that the stairs be used.

3. If the fire hat symbol is flashing in the car, be aware that shunt trip (see chapter 22) may end your trip up the building without further warning. You may end up marooned between floors.

4. Members must be careful during any emergency stop. Its use must be regarded as an emergency action, because the car may not start again after an abrupt and jolting stop.

5. In new installations after 2004, an emergency stop switch is available in the *fire operation panel* while on Phase II. Since 1987, they had not been installed in passenger elevators. Check with your local AHJ about your situation.

6. If moving in smoke-filled floors, always follow the rule of advancing a tool along the floor ahead of you. A number of firefighters have been killed or seriously injured from falls down hoistways due to hoistway doors missing or open due to the fire conditions.

Effects of Fire on Elevator Components

Fig. 24–7. Destroyed by fire (photo by Ed Fowler, CFD)

Due to the fire environment, the mechanical or electrical systems can be affected by heat or water causing erratic and dangerous behavior of the elevator (fig. 24–7). In the history of building fires, floor doors have been ripped off their mountings as an elevator slid away from a floor during a fire. Firefighter fatalities resulted from this opening being left unprotected and open in a smoke and heat filled upper lobby of a fire building. Hoistway doors have been warped beyond function by the extreme heat of a fire floor, causing interlocks to be destroyed (fig. 24–8). Hoisting ropes, which are made of wire steel, have melted and separated due to the heat they were exposed to during a fire (fig. 24–9).

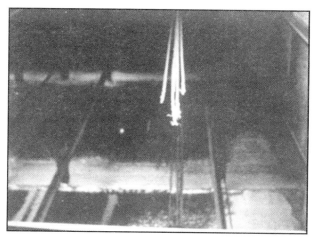

Fig. 24–9. Separated hoist ropes from elevator at the MGM Casino fire (11/21/80) (photo courtesy of Ed Donoghue Assoc.)

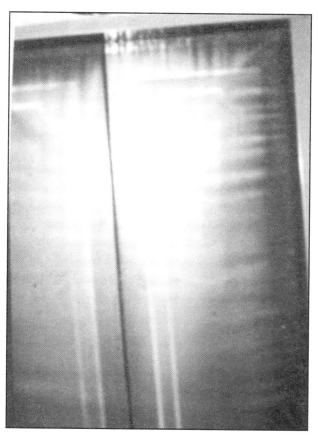

Fig. 24–8. Smoke marking on doors (photo courtesy of Ed Donoghue Assoc.)

Water from broken pipes, sprinklers, and hand lines can cause electrical short circuits that may disable the cars, making their use impossible. In any situation where the elevator operates erratically, exit the elevator at the nearest floor immediately. Place the car out of service, and immobilize it if possible. Remember, when elevators are erratic, they can and have operated with their doors open, moving up and down the hoistway at will. All bets are off—the fire environment has won this elevator!

Firefighters Trapped in an Elevator During Fire Operations

There can be no more frightening experience for a firefighter than being trapped in a stalled elevator at or near the fire floor. When we think of our fear of being lost in a cellar fire or some other harrowing situation, it pales to the fear this must bring out. Many firefighters have lost their lives in this situation, or had the lifelong memory etched into their psyche. The elevator may have stopped for any number of reasons, including shunt trip (see chapter 22) activation or fire damage conditions to the system.

1. If during the operation, the elevator containing members stops on a floor with smoke or fire, attempt to close the car door, using force if necessary.

2. Select another floor to get the car away from the fire floor.

3. Turn the Phase II key switch to *off,* and hopefully the car will be recaptured by Phase I Recall.

4. Check the door safety circuit contact by pushing the door closed again.

5. Check position of the emergency stop switch, if provided, and move it to *run.* Enter another floor selection.

6. Check the position of the Phase II key switch; make sure it is in the *on* or *run* position.

7. Communicate your situation to the IC. Make sure that the location that you are giving is the correct one. Firefighters have died after being lost on floors of high-rise buildings, while search parties looked for them on the wrong floor location that they had given to the IC. (One Meridian Plaza, Philadelphia. February 23, 1991)

8. If you are going to attempt a self-rescue, notify the IC immediately of your plan and the location, number, and condition of the members.

9. Put all gear on to ensure the maximum level of protection.

10. Using the knowledge of the floor plan, force the car door open and exit onto the floor if conditions will allow. Stay together by holding another members foot or ankle as you crawl and follow each other out to the stairwell. *Stay with the wall!*

11. Take a head count, and notify command of your successful self-extrication of the members!

12. If the normal exit is not usable due to fire conditions or if a restrictor is in place (see chapter 4) then exit via the top emergency exit.

Fig. 24–10. Forcible exit from elevator via top emergency exit

13. The top emergency panel will have to be forced open with forcible exit tools, but it should be easily removed (fig. 24–10). As the first member exits onto the top of the car, he or she must engage the *stop switch* located on the inspection box on the crosshead beam (fig. 24–11). There is also a working light switch there to assist with vision.

14. Locate the nearest floor door and trip the interlock to exit onto the floor. Secure the door after all have exited onto the landing, and notify the IC of your situation.

Fig. 24–11. Car top inspection box with stop toggle and light switch

Fig. 24–12. Fully equipped firefighters awaiting elevators on Phase I recall (photos courtesy of Department of Fire Services, Massachusetts)

At this point, one's own department procedures will dictate what now takes place. The listing of what should be carried by the members changes with the borders of the individual community, or, unfortunately, with the likes or dislikes of the battalion chief *working that day* or the company captain. This is where an SOG comes into play and sets the rules that everyone must follow (fig. 24–12). The minimum *must* consist of the following to provide a means of forcible *entry* or forcible *exit* from the elevator in needed:

- Flat head axe and Halligan bar

- 6-foot plaster hook

- Hydraulic ram tool

- Engine company bags with the selected lengths of hand line hose, nozzles, and the like, for interior firefighting

It must be emphasized and recognized that the SOG equipment load can and must be adjusted to the staffing ability at the local level. What can't go up with the first companies in the elevators must be sent up as soon as possible.

Fig. 24–13. Always check hallway floor plan.

After checking the floor plan schematic *(you are here!)* that is usually on the wall, locate the apartment or section of the building that you are en route to (fig. 24–13). Check this plan on the *first* and *last* floors that you stopped at, because floor plans can change significantly from floor to floor. By finding that location, you can then locate the closest stairwell to your desired destination on the fire floor. This is critical when running hand lines off of the standpipe connection on the floor below or above. Every engine company member who has worked in the apartment or high-rise building setting has at one time or another ended up with a hand line that was short and wouldn't make the desired apartment's door.

At this point, we would direct you to the many excellent books dealing with interior fire attack, high-rise firefighting, or any of the other publications that deal with the subject *after* you have gained entry to the fire floor.

Summary

As we have stressed, the use of elevators in a burning building is a risky proposition. Only by being properly trained and equipped can we hope to draw an even chance in these situations. Elevators and fire conditions do not mix well, but by using the fire service function in the elevators, we can escape some of the hazards that we know exist in the fire building.

All members must be familiar with firefighters' service operation, Phase I and Phase II, before they ever step into an elevator to travel up into a building to fight a fire. This is not an on-the-job training type of function, but must be done under no-stress conditions during training exercises. Remember and enforce the rule of six:

- Do not use the elevator if fire is on sixth floor or below.

- No more than six fully equipped firefighters in the car.

There may be any number of variations in your own response district due to the improvements in the system as years have progressed. The fire service must get out of the stations and into its buildings to know what they are going to face at 0345 hours.

Did the elevator respond to the lobby on automatic recall?

Proper tools and equipment listed previously must be carried in with any fire company entering an elevator. They may be necessary to perform *forcible exit* as well as *forcible entry*! All firefighters must familiarize themselves with the location of the top emergency exit in elevators.

The vital importance of having and following an SOG is critical to both the life safety of the members and the overall success of the response.

Finally, it is equally important to review the sections we covered:

- Precautions to the use of elevators during a fire

- Effects of Fire on Elevator Components

- Firefighters Trapped in an Elevator During Fire Operations

We hope we made the point that there are extreme hazards when entering a building and using its elevators during a fire. The attitude of "We've been here a hundred times," causing members to let their guard down, has cost firefighters their lives. Don't become the next one that we must read about! And even more emphatically, do not become the next one whose death was caused by someone else's poor decision. There are two pictures of this elevator (fig. 24–14) in our possession, one showing five bodies of the unfortunate occupants, with their luggage, who died in this car that fateful morning at the MGM hotel. As firefighters, we always pay respect to those who die in a fire, so we do not show those victims. Rather we show the outline of their bodies, left by the smoke that killed them.

Fig. 24–14. Outline of victims who died in MGM Grand Hotel elevator (photo courtesy of Ed Donoghue Associates)

Review Questions

1. What version of ASME A17.1/CSAB44 is currently in use in your state or region?

2. Do you know the operational aspects of automatic recall?

3. What is the rule of six?

4. Describe when and where you will find an emergency stop switch/button.

5. What does it mean when the fire hat symbol is flashing?

Field Exercise

After determining which version of the ASME A17.1 code is being used in your area, conduct five field visits to high-rise apartment buildings in your first-alarm response. List those that have automatic recall and alternate floor capability.

Endnotes

[1] "Multiple Fatality High-Rise Condominium Fire." Clearwater, Florida. USFA-TR-148/June 2002.

Chapter 25
Training for Safe and Successful Elevator Rescue Operations

Introduction

In large city or county fire departments, firefighters usually respond to elevator emergencies on a daily basis. Yet, other fire departments might respond to only a handful of elevator emergencies annually. Of course, the more elevator incident responses, the more opportunities firefighters will have to gain experience in removing passengers entrapped in stalled elevators.

Some fire departments assign the day-to-day responsibility for responding to elevator emergencies to one of their specialty teams such as special operations or technical rescue, or to a ladder company or rescue squad. The benefits of limiting a relatively small number of emergency units and firefighters to respond to this type of incident response are that (1) it provides for specialized and dedicated training on the subject of elevator rescue, (2) it allows the firefighters to gain expertise and build teamwork, and (3) it allows them to become more familiar with the elevators within their response area.

In other fire departments, elevator incidents are managed on a first-due basis. The closest unit(s) to the incident responds. Using this approach may work well

in a fire department with a small number of elevators, and when the elevator mechanic is "just around the corner." Yet, training is still recommended to ensure that firefighters have the basic skills and knowledge to safely manage elevator incidents.

Regardless of the size of a fire department or its composition (e.g., career, volunteer, or combination), the number of elevators in its response area, or its elevator incident response history, firefighters should rely on a combination of training and experience to acquire an effective response capability.

Experience alone does not necessarily mean that firefighters will be following sound safety principles and good passenger evacuation guidelines. Though well intentioned, firefighters can become victims of an elevator rescue event if they fall short of safety objectives. This can result from failing to remove the main power from an elevator, unauthorized and premature restoring of power, entering or working near an exposed hoistway without using fall protection equipment, or implementing unsafe passenger removal procedures.

Training is the foundation of safe and effective elevator rescue operations. Training and field experience are the means by which firefighters gain both proficiency and confidence in the management of elevator rescue incidents. Elevator rescue training introduces new firefighters to the basics of how an elevator operates, terminology, hazards, safety, common elevator rescue tools, and basic passenger removal guidelines. The training also serves as a review, refresher, or means to introduce new or specialized guidelines to firefighters who already have knowledge above the basic level.

Training Levels

Although we recognize that there are many ways to approach elevator rescue training, we believe that training officers should consider developing elevator-training modules. Training officers and other qualified instructors can deliver these modules in a progressive sequence. That is, each successive module should build on the scope and objectives of the previous module. Moreover, the training contact hours and training environment will likely differ depending on the particular module. For example, one module might require three to six hours, all delivered in a classroom setting. Another module might take place during the same period, as a combination of classroom and hands-on training. Still another module might require the same or longer time commitment and consist of hands-on training only.

Training officers should consider many factors when deciding which approach is the most practical and economical to meet their local training needs. The local elevator response history is a major factor for consideration in the decision-making process. Some other factors are:

- Scope of training
- Length of training sessions
- Training environment (i.e., classroom, elevator building, etc.)
- Required tools and equipment
- Target audience (e.g., entry level firefighters, special operations team)
- Refresher training (type and number of hours)
- Program evaluation process
- Funding (e.g., overtime, administrative, and equipment costs)

There is no one training approach that meets the needs of all fire departments. Yet, training officers are encouraged to contact neighboring fire departments, their state fire training organization, and the local elevator company to gain information and insight about how they manage elevator rescue incidents. Although there will probably be differences among these resources, there likely will be similarities as well. Select what meets your needs and prepare your own training guidelines and program.

The design, contents, and scope of these training modules can serve as the basis for elevator response readiness levels. For discussion purposes, listed next are five training modules. These generic modules provide training officers with content building blocks on which to develop local training requirements. With the exception of the first module, to participate in the other modules, the participants must have successfully completed all of the previous modules. That is, the previous module is the prerequisite training for the next module. Training officers are reminded that these modules are provided to stimulate interest and evoke action only.

Module 1: Operation of firefighters' service (Phase I and Phase II operations) and standard operating guidelines (SOGs) for the use of elevators in high-rise buildings under fire conditions.

Module 2: Basic components and terminology of an elevator, operation of hoistway door unlocking device keys (e.g., drop, lunar, and tee), poling guidelines (poling across and down), safety principles and rules, passenger removal guidelines where an elevator is stalled at or near a landing or within three feet of a landing, and management of an incident involving a victim that fell down the hoistway.

Module 3: Passenger removal when the elevator car is stalled more than three feet above or below a landing, rescue from a blind hoistway, and use of forcible entry. This operation may require removing passengers through the top emergency exit, breaching a wall, or rappelling.

Module 4: Person pinned between car and hoistway wall, counterweight, or other equipment. This situation requires the development of special standard operating guidelines and the combined resourcefulness of special operations or technical rescue teams and the local elevator mechanic.

Module 5: Rescue of firefighters or passengers trapped in an elevator during fire in a building. This situation warrants the development of special operating guidelines and the combined resourcefulness of special operation or technical rescue teams and the elevator mechanic.

Training Setting

Training academy

Some fire departments include elevator rescue training as part of their in-service training program. Crews of firefighters are scheduled to attend either while on or off duty. In the former situation, firefighters take their emergency vehicles to the training academy (center) as well. This allows crews to be available for emergency response. However, when firefighters must respond to incidents during the training session, it disrupts the learning process and, as a result, often requires rescheduling of the session. The disruption also can add an unplanned administrative and financial burden.

Recruit Training

Including elevator rescue training as part of the recruit-training program is one way to provide basics of elevator rescue operations to the recruits. We recommend at least a 6- to 12-hour block of time for the training. Preferably, the training should be a combination of classroom lecture/discussion and hands-on activities. The training should include topics such as:

- Basic elevator terminology, tools and equipment

- Types of nonemergency and emergency rescue operations

- Phase I and Phase II operations of firefighters' service

- Safety principles and rules

- Operation of hoistway door unlocking device keys

To help measure the effectiveness of the elevator rescue training, the training should be part of the recruits' evaluation process. If the training is not part of the required evaluation process, recruits are not as likely to review the information with the same level of commitment as for the "required" subjects.

In-Station and Area Training

Providing training in the station or in the local response area is okay, as long as it is not a busy station. That's a challenge for some departments. A possible solution is to use the SOGs as a means of review. In that way, if the training gets disrupted and not everyone must respond, the firefighters that remain in the station can review the guidelines on their own.

Stations or departments with a low response demand tend to have more time to devote to elevator training. That is a good thing when one stops to consider that training is a form of positive reinforcement of skills that are infrequently used. We do recognize and appreciate that not all fire departments have elevators in their local response area. However, if neighboring jurisdictions do have elevators, it would be wise for firefighters to learn how to manage elevator incidents in the event they are requested to respond. This can result because of a serious fire or a major power outage in the neighboring area. It's a nice feeling to know that when firefighters respond to elevator emergencies either in or outside their immediate response area, they are prepared. No one wants to respond to any incident if they are not confident or trained to manage it. En route to an incident is not the time to realize that you may or may not be prepared to do what may be expected of you.

State Training Programs

Many state fire and rescue organizations offer elevator rescue training. Two excellent examples are the Massachusetts Firefighting Academy (MFA) and the Maryland Fire and Rescue Institute (MFRI). Each organization offers elevator training at their training facilities and also in outreach programs. Firefighters get the opportunity to learn about how an elevator operates, identify the four safety principles, and learn ways to safely remove passengers from stalled elevators. The training usually consists of classroom and hands-on training.

The Out-of-Town Instructor

The fire service has many good elevator rescue instructors throughout the country. Most of them are still active with their fire departments and serve as members of the instructor team. Other firefighters provide training to other fire departments on a part-time basis. And others have retired or otherwise left the service and become elevator rescue training instructors or consultants. In any case, they might have differences in their tactical approach. Still, they have the same concern for the safety of firefighters and entrapped passengers.

We lightheartedly refer to the instructors who provide training to other fire departments as the "out-of-town" instructors. This is merely a reference to their need to travel, not their level of expertise. Differences aren't bad as long as the safety principles remain the same. Elevator rescue training essentially is about safety, problem solving, and proper use of tools and equipment.

Training Center Elevator Props

Some fire departments have acquired elevator cars as donations from contractors who are demolishing buildings. Other fire departments might be thinking about acquiring an elevator. On the surface this idea has merits. Yet, before committing to acquiring an elevator for a static training prop, the training officer and other fire officials should first ensure the following:

- The training benefits of the elevator car warrant its acquisition.

- There is sufficient space at the training facility to install the elevator car.

- There are sufficient means and resources needed to transport and place the elevator car.

- There is enough assistance available from the local elevator company in setting up the prop(s).

- There are no hidden legal or liability issues related to acquiring the elevator car from the building demolition company or other responsible parties.

If practical to achieve, the elevator car should have its car door power operator still attached, and the hoistway door and doorframe assembly available for reassembly at the training center. Requesting assistance from the local elevator company is key to maximizing the utility of the training prop. Depending on available resources from the elevator company and the level of commitment of the fire officials, the elevator car training prop can offer the following benefits:

- Familiarization of the car operating panel (COP) and other common elevator equipment

- View operation of hoistway door key and interlock from inside and outside the car

- Use of hoistway door unlocking device key

- Use of poling to unlock the hoistway door from a platform constructed above or along side of the hoistway door

- Familiarization of how the hoistway and car doors operate

- Simulation of passenger evacuation through the car's top emergency exit

- Use of the car door operator assembly to manually open the doors

The training officer can enhance elevator-training scenarios by setting up a simulated (dummy) elevator mainline power disconnect switch near the elevator car prop. Students can practice removing the power from the stalled prop elevator car and using lockout/tagout procedures. The more practice in using these activities, the better firefighters should do during actual elevator emergencies.

Other useful training props are cut-away versions of the hoistway door unlocking devices and keyholes that allow use of the appropriate hoistway door unlocking device key. The local elevator mechanic is an excellent resource to consult about how to obtain and construct these props. Having props that show and allow operations of the drop, lunar, and tee hoistway door unlocking device keys would enable easy transport by car and use in fire stations.

Preparation for Training

Before scheduling the training, the training officer should use a checklist to ensure that the planned training has clearly defined objectives, the activities are identified, and the firefighters are informed of the specific activities and their required level of participation. The checklist should include points such as:

- Type of training (e.g., classroom, hands-on, etc.)
- Use of activity stations and participant rotation schedule
- Schedule for breaks, rehydration, and lunch, if applicable
- Location and duration of the training
- Specific training activities or procedures
- Elevator tools and equipment required
- Personal protective equipment required
- Need for lightning to illuminate the rescue area, if necessary
- Approval to use a building for hands-on training
- Contingency plan if the training is disrupted because of need to respond to emergency incidents
- Inviting an elevator mechanic to assist with the training

The success of the training activity largely will rest on how well the activity is organized and planned, the level of preparedness and experience of the instructors, and how well the students are ready to participate. If the seasoned firefighters perceive the training to be old hat, poorly organized, or unchallenging, they may become disinterested or distracted. This may reduce the effectiveness of the training.

When the training session is aimed at orienting new firefighters on the basics of elevator rescue operations, and there is a mix of experience levels in the training session, the instructor(s) should use the firefighters with more elevator rescue experience to help the less experienced firefighters.

The Instructor Cadre

An elevator rescue instructor may be a firefighter who specializes in elevator rescue operations, a firefighter with experience as an elevator mechanic, or a local elevator mechanic. These individuals are essential to the success of elevator rescue training programs. They are the ones who share their practical wisdom and guidance to help firefighters become more knowledgeable and capable to manage elevator incidents.

Elevator mechanics are among the many experts in the elevator industry. No one knows the ins and outs of an elevator better than they do. Although it is common to have an elevator mechanic respond to an elevator rescue incident either at the request of the on-scene rescue team leader or building representative, this does not necessarily mean that the mechanic also will be present during elevator training. You need to invite the mechanic(s). Usually, all it takes is a phone call from the training officer or other designated person. It's been our experience that if you invite them, and their schedule allows, they'll be glad to attend and share information.

However, some elevator mechanics do not participate because they are not allowed to share information with emergency responders. Although they are willing to help, their company policy may prohibit them because of liability reasons. If that is the case, firefighters should respect their position.

Throughout our respective careers, we have met many outstanding elevator mechanics, inspectors, and members of elevator management teams whom have made tremendous contributions to our elevator rescue training programs. Sometimes the training is provided on an informal and impromptu basis, such as at the site where they are installing or repairing equipment. At other times, information is shared during and immediately following an elevator rescue incident. Still other times, the elevator representative shares information in an often more formal setting such as a classroom or at the site of elevator hands-on activities.

Use of Elevator Buildings

An elevator in a parking garage or building serves as an excellent training venue for elevator rescue training. However, before using the elevator, firefighters must contact and gain approval from the owner or property manager. After all, the building owner owns the elevators!

The owner or property manager's main concerns are elevator out-of-service time, inconvenience to occupants, safety, potential damage to equipment, and liability. For these reasons, it is important to let them know the purpose and particulars of the training activity, safety precautions, and requested date(s) and time of use.

Elevator Mechanic Involvement

It makes good sense for the elevator mechanic who services the elevator(s) to be present during the training activity. The training officer should contact the elevator mechanic before scheduling the training session to confirm availability. During the conversation, the training officer also should provide the elevator mechanic with an overview of the training and what particular assistance is needed. Some buildings, such as large hotels, have an elevator mechanic on duty during the day, five or six days a week. This is a win-win situation for the building management, employees, and visitors. It also is a plus for the local firefighters should the elevator incident require their services.

The presence of an elevator mechanic during the training session offers four distinct benefits to firefighters engaged in the training. The first is resourcefulness. Technical questions often arise during training that are better answered by an elevator mechanic. Second, the elevator mechanic can add information peculiar to the elevator(s) used in training.

A third benefit is that the elevator mechanic is available to immediately restore the elevator should it stall. We know very well that it is not uncommon to have an elevator stall during training. Another benefit is that the elevator mechanic can not only share safety tips, but can also help the safety officer in monitoring the training. During our respective fire service careers in Cambridge, Massachusetts, and Montgomery County, Maryland, we participated in a number of hands-on elevator training sessions where the training elevator failed to operate during its use.

When this happens, an undesirable situation quickly develops both for instructors and students. If the elevator fails to run after the training session has ended, that's one thing. For it to not operate during training, that's the least desirable situation of the two. However, if the elevator mechanic is present, then usually that problem is quickly fixed and training can resume. If the training officer is not so fortunate as to have the resources of an elevator mechanic at the training site, the impact can be the loss of training time, need to reschedule the training, and possible disappointment and dissatisfaction among the firefighters participating in the training. This is an often hidden concern that later becomes apparent when the training officer visits the management representative before leaving the building. It might be awkward and embarrassing to the training officer when he or she meets with the building representative following the training. After thanking the representative, the training officer might add, "Oh, by the way, you'll need to call the elevator mechanic because the elevator doesn't work. I don't know what happened! Have a nice day!"

Safety

If you don't adhere to safety principles and rules during training, how do you know you will during an actual elevator rescue event? As all of us know, the training environment simply should be one step removed from reality. The major differences are that the training venue is a controlled environment and the other is not; firefighters are not under the stress of uncertainty and the sense of urgency to evacuate a passenger or extricate a pinned victim; and no one is at any risk of harm except for the firefighters themselves.

The safety regimen must be to power down the stalled elevator and any others that might be a safety issue; use lockout/tagout procedures; and wear head, eye, hand, and foot protection. And if the training activity should require working in or near the hoistway, then fall protection is also required. Another concern is protecting the passengers. If the training activity is a simulation of rescuing a passenger from a stalled elevator through the elevator car's top emergency exit, then the firefighter who plays the role of the passenger must also be protected appropriately. Exercising SOGs should identify areas that are valid, and also any that require change.

Also, be sure to assign a safety officer to oversee the operations. It's easy for firefighters to get so caught up in the learning experience that they sometimes lose sight of hazards around them. This is especially true when there are several firefighters engaged in the same activity in proximity to each other.

Standard Operating Guidelines

Elevator rescue SOGs are designed to streamline operations, eliminate confusion, promote safety, and identify firefighters' duties and responsibilities. Elevator rescue training activities not only serve to develop or improve proficiency and build confidence among firefighters but also help to validate the utility and effectiveness of the guidelines. Exercising SOGs should identify areas that are valid and also any that require change.

Don't forget to get input from the local elevator mechanic. This extra review and constructive feedback might reveal the need to adjust rescue tactics, reevaluate safety measures, or address potential liability concerns.

Participation

When conducting hands-on training such as using hoistway door unlocking device keys, poling procedures, or simulated passenger removal through an elevator car's top emergency exit, the instructors should ensure that all attendees participate in each activity. To help ensure the efficient use of the training time to meet training objectives, the instructors should divide the training into different workstations. The number of stations will depend on the available time, number of participants, and availability of elevators. For example, there can be a station located in the elevator machine or pump room, for equipment familiarization. At this station, the instructor (firefighter or elevator mechanic) can explain different components of an elevator such as the hoist machine and ropes, safety governor, controller, and main power disconnect.

Another station should be set up for the operation of a hoistway door unlocking device key. Each firefighter should have an opportunity to actually operate the hoistway door unlocking device key and

view its operation from the hoistway side as another firefighter is using the key. Using a poling tool can be another station.

Learning stations are usually effective as long as there are not too many participants for the number of stations. When this is the case, the wait time for idle firefighters often leads to their dissatisfaction and disinterest in the training. On the other hand, if there are too many stations, then the learning time per station will not be adequate. Striking a good balance often comes by trial and error. When planning the stations, be sure to give consideration to actual training time per station, time to rotate groups, and break time.

Firefighter Elevator Training Record

Fire departments usually maintain hard-copy or electronic records of the training delivered to and received by its members. Using an electronic management information system allows the training officer to quickly access and analyze the elevator training history by individual, station, battalion, or the entire department. Having access to the department's elevator response history, including frequency, types, and distribution of incidents, and station and battalion response profiles, can be beneficial to the training officer when developing new and refresher elevator training. Furthermore, data mining to the individual level should provide the training officer with a better assessment of each firefighter's elevator rescue training level, individual response frequency, and need for refresher training.

An elevator training and response information storage and retrieval system can play a vital role in ensuring that firefighters are trained, prepared, and confident in the principles and practices of safe and effective management of elevator incidents.

In many fire departments, elevator incidents are few and far between. Still, firefighters must be ready to respond to a stalled elevator with passenger entrapment or even worse. If the frequency of elevator incidents for a fire department is relatively low, then the frequency of elevator rescue training should be relatively high. Conversely, if firefighters respond to a high number of elevator incidents and they have good standard operating guidelines, then refresher training can be less frequent.

Generally, firefighters should train the most on subjects, of equal importance, that they do the least. An exception to this statement is the case where there is a demonstrated need to train on a frequently used procedure for one or more of the following reasons: The procedure has changed, a firefighter(s) had difficulty safely executing the procedure, or there is a firefighter who is new to the procedure.

Building Survey Program

Just as for other types of incidents, the more firefighters know about the elevators in their response area, the better prepared they will be when they respond to manage incidents involving them. Although elevator equipment from one elevator manufacturer or company to another is very similar in operation, there still are differences. If the number of elevators in the local response area is a manageable number to survey, firefighters should prepare a schedule to visit each of them to collect information. The purpose of conducting the survey is to gather useful information that would be beneficial during an elevator incident such as:

- Name and address of the building

- Number, location, and types of elevators (electric traction, hydraulic, passenger or freight)

- Standby power system

- Elevators under standby power system

- Whether the elevator system is equipped with a shunt trip

- Types and location of escutcheon plates

- Types and location of hoistway door unlocking device keys

- Location of the elevator machine room (electric traction) or pump room (hydraulic)

- Firefighters' service, Phase I and Phase II operations

- Whether poling across or poling down is usable

- Type of door restrictor and method to release

Entering the collected information into a database program enables firefighters to prepare useful reports about the elevators in their response area. For example, knowing the number of elevators would give firefighters an indication of the potential number of stalled elevator incidents resulting from a power blackout in all or part of the response area. As a result, firefighters can estimate the resources needed to manage the number of elevator incidents following power loss.

Summary

Elevator rescue training is needed to help firefighters build both confidence and proficiency in the safe management of elevator incidents. Training should include both classroom and hands-on activities.

One approach to delivering elevator information is to develop training modules. The modules can run from simple to complex. In this chapter we suggested five modules and provided examples of the kinds of information included in each module. It is left to the firefighters to determine what information is provided, the target audience, and the required number of hours to complete.

The training modules can lead to corresponding response levels. For example, firefighters who successfully complete Module 1 would be trained in elevator response level 1. Completion of each module would result in a corresponding elevator response level.

Knowing who has what level of elevator rescue training can be useful when there is a call for a stalled elevator with passenger entrapment. The officer can check quickly to determine if the firefighters present are trained to safely manage that type of incident.

The success of the training program rests with the quality of the training material, the quality of instructors, and the readiness of the participants to learn. The training consists of classroom and hands-on activities. The productivity of the training is also influenced by the presence and resourcefulness of the local elevator mechanic.

Refresher training is necessary to help maintain skill level and to introduce new information. Regardless of the type of training, safety must be an integral part. Remember and follow these principles: Remove the main power from the training elevator (and other elevators if involved), use lockout/tagout procedures, guard hoistway opening, protect firefighters, and protect passengers. Be ready to train and train to be ready!

Review Questions

1. Why should firefighters be trained in elevator rescue operations, even though there are no elevators in their first due response area?

2. When planning an elevator training session, it is useful to have a checklist of items to consider. In this chapter, we mentioned 11 items. List eight.

3. List three of the benefits of having an elevator mechanic present during an elevator training session.

4. Elevator rescue SOGs are designed to streamline operations and eliminate confusion. List two other benefits.

Field Exercise

Visit a building to check the suitability and limitations of using one or more elevators for elevator rescue training. Call and meet with the building manager or engineer for approval before inspecting the elevators.

Chapter 26
Escalator Development History

History

The first escalator was built in 1896 when Jesse Reno made a little 6-foot stairway that lifted people onto the Coney Island pier in New York City. Four years later, the 1900 Paris Exhibition displayed four different kinds of escalators, including the Reno. The Coney Island success brought Reno more business, including installations in major department stores and the subway systems of both New York City and Boston (fig. 26–1).

Another participant at the Paris Exhibition in 1900 was Charles D. Seeberger, who had joined Otis Elevator in 1899. He is credited with coining the word *escalator,* (which was created by joining scala, which is Latin for steps, with elevator). Early escalators were also known by a variety of names, including traveling staircase, inclined elevator, and magic stairway. The Seeberger-Otis union produced the first step-type escalator made for public use, and it was installed at the Paris Exhibition, where it won first prize.

The spread of the installation of escalators was slow at first, because there weren't many high-rise buildings at that time. Later, as more and more high-rises went up, the demand for escalators significantly increased. The

Fig. 26–1. The Reno escalator (courtesy of Otis Elevator)

need for an escalator was driven by the need to transport large numbers of passengers up or down a slope of 30°. Finally, the competition between the companies and designers ended up with a unit that could transport 8,000 people an hour up a 30° slope successfully. The real role of escalators is not so much that of labor savers but of space savers. They keep people moving in crowded places, where it is desirable to not have people congregating, possibly blocking passenger flow.

Modern Escalators

The use and installation of escalators has progressed continually with better designs and safety systems installed in accordance with ASME A17.1 Safety Code for Elevators and Escalators. Early escalator steps were fabricated of steel and wood, or metal treads fastened with screw bolts. Today, escalator steps are usually constructed of aluminum. In some instances they are fabricated from several pieces, but some manufacturers are currently casting them as one piece. In the succeeding chapters, we will review and discuss their basic design, improvements, accident history, and injuries and deaths associated with this type of equipment.

One only has to look around as we travel to see the role that escalators play in moving large numbers of people. Places of assembly, airports, transit systems, sports facilities, malls, and the like are suitable places for installation of these units. A problem area that has developed along with the increased use of escalators is the incorrect use of the system by passengers. Baby carriages, carts, two-wheelers being used making deliveries, and wheelchairs are all found using the escalator. These actions are in part what cause damage to the *combplate* and *steps*. Part of the overall access plan for the building or mall will be the adjacent installation of an elevator to address those needs. A real eye-opener for anyone is to watch the actions of people and their children on an escalator. The next time you visit a mall, take a few minutes to observe the human interaction with the escalator from a high vantage point. It will leave a person breathless!

As firefighters, we are called often to incidents with entrapment, resulting in injuries, and occasionally ones resulting in death. In the succeeding chapters, we hope to give you the tools and information to assist you with this type of incident.

The escalator of today is a far cry from the Reno and others of its generation. The Reno was only wide enough for one passenger per step. The steps were made of wood, and they were very uncomfortable to stand on. The noise the Reno made assured that a passenger was not lulled into a trance, but rather it caused you to pay constant attention to where you were in the ride. Finally, the wooden steps ejected the passenger at the landing two inches in the air, not coasting into a smooth transition step flush with the landing as is done today. Although the escalators of old were noisy, uncomfortable, and ugly to look at, the improvements to the system are part of the problem.

Today's new escalators are wide enough to carry two passengers side by side. They are very quiet, comfortable to stand on, and smoothly arrive at the landing and transition into the landing step. That is part of the overall problem, and ASME, the escalator industry, the Elevator/Escalator Safety Foundation (EESF), as well as individuals like Carl White, of Carl White Associates, were leading early proponents of making the escalator ride a safer one. The EESF was started in 1989 and put its first program forward in 1990, called Safety Rider. In the following chapters, we will explore the escalator itself, as well as the role of the EESF and the newer designs aimed at ending the deaths and injuries that are associated with escalators.

Escalator Fires and Smoke Spread

The occurrence of fires involving escalator systems has been reported many times in the past, sometimes resulting in loss of life. This brought about code changes in both the United States and Europe to protect the riding public from the spread of fire and smoke via these systems. As they are used for the transportation of the public in transit stations, retail stores, malls, sporting facilities, and many other venues, there can be a significant threat to the life and safety of those riding on them.

Part of the problem in the past was the collection of debris under the truss construction and other areas of the machine that could become fuel for a fire. This was mixed with the oil and grease that are part of an escalator drive system. The fire could be the result of malicious arson, accidental ignition from discarded smoking materials, or any number of other reasons. The result is the ignition of whatever fuel is there and the upward spread of the smoke and heat via the channel that the escalator itself creates. Whether this situation is in a subway, department store, or a mall, it can result in many injuries, deaths, and the uncontrolled spread of fire. The Kings Cross fire in the London transit system in 1987, where 31 people were killed, is a prime example of such a hazard.

In our modern installation, various codes mandate the use of automatic sprinkler protection and smoke detector–controlled rolling shutters to prevent fire extension or smoke spread into the floor above.

Summary

The escalator as we know it has been around since 1898, when Jesse Reno made his presentation at the 1900 Paris Exhibition. Another inventor of that era was Charles D. Seeberger, who is credited with creating the word *escalator*. They have also been called by other names, such as traveling staircase, inclined elevator, and magic stairway.

The escalator is intended to carry passengers up a 30° slope, with a plus or minus of no more than 1°. Due to a variety of reasons and causes, they have been the site of a number of severe and sometimes fatal accidents involving passengers. The Elevator/Escalator Safety Foundation (EESF) is a leading champion of teaching safety practices to children and elderly adults, both groups that suffer most of these entrapments.

Escalators are covered by the ASME A17.1/CSA B44 Safety Code for Elevators and Escalators-2004, section 6.1.

Review Questions

1. What two words were used to create the word *escalator*?

2. What did Jesse Reno invent?

3. What is one other name that was used for an escalator?

4. Who was Charles D. Seeberger?

5. What does EESF stand for, and what is its function?

Field Exercise

Visit a mall or transit station in your inspection district and familiarize yourself with the escalators installed there.

Chapter 27
The Escalator System

Escalators are some of the largest, most expensive machines people use on a regular basis, but they are also basically very simple, consisting of a pair of rotating chain loops that pull a series of stairs in a constant cycle, moving a lot of people a short distance at a good speed. The maximum angle allowed is 30°, plus or minus 1° due to field conditions during installation. An escalator moving 100 feet per minute (ft/min) is the standard according to ASME A17.1-2000. At that speed, they are capable of delivering many more passengers over the same distance than a standard elevator. The typical unit weighs six tons and is designed to handle at least 3,000 passengers an hour. In a typical building open for 10 hours a day, an average escalator can handle 30,000 people. It should be noted that this is a statistical figure, given that there are usually not that many people involved. According to an article by Ray Lapierre of the Elevator and Escalator Safety Foundation that appeared in *Skylines* magazine in 1997, an estimated 33,000 escalators in the United States and Canada move 245 million escalator passengers daily (fig. 27–1).

Escalator Components

Fig 27–1. The escalator system

The escalator system consists of the following components:

- The *landing:* The floor plates are level with the finished floor and are either hinged or removable to permit access to the machine spaces under them (fig. 27–2).

Fig. 27–2. The landing

- The *combplate* is the piece between the stationary landing and the moving step. It slants down slightly so that the comb teeth fit between the cleats on the steps (fig. 27–3).

Fig 27–3. Combplate interface with step

- The front edges of the comb teeth are below the surface of the cleats. The comb is usually six inches in length, and the teeth are fastened in a row with screws (fig. 27–4).

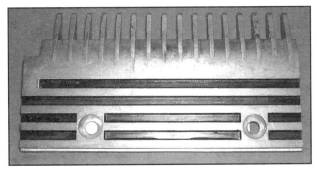

Fig. 27–4. Section of combplate

- The *truss:* The truss is the mechanical structure (the *frame*) that bridges the space between the lower and upper landings. The truss is basically a hollow box made up of two side sections joined together with transverse braces across the bottom and just below the top. The ends of the truss rest on concrete or steel supports (see fig. 27–9).

- The *tracks:* The track system is built into the truss to guide the step chain, which pulls the steps through an endless loop. There are two tracks: one for the front of the step (*step-wheel track*) and one for the trailer wheel of the step (*trailer-wheel track*). The relative position of these tracks causes the steps to appear from under the combplate to form a staircase, then disappear back into the truss.

The *reversal track* at the upper landing rolls the steps around the top and starts them back in the opposite direction. An overhead track ensures that the trailer wheels remain in place as the step chain is turned back on itself (fig. 27–5).

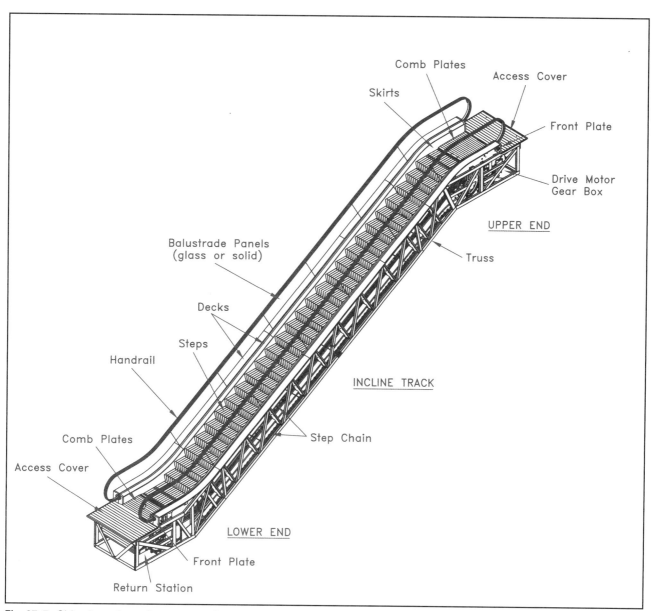

Fig. 27–5. Side view of escalator system (courtesy of KONE Corp.)

- The *step:* Each step in the escalator has two sets of wheels, which roll along two separate tracks. The upper set (the wheels near the top of the step) are connected to the rotating chains, and so are pulled by the drive gear at the top of the escalator (figs. 27–6 and 27–7).

Fig. 27–6. Step turning under

Fig 27–7. Front view of steps going under

- The other set of wheels simply guides along its track, following behind the first set (fig. 27–8).

Fig 27–8. The tracks are spaced apart in such a way that each step will always remain level. At the top and bottom of the escalator, the tracks level off to a horizontal position, flattening the stairway. Each step has a series of grooves in it, so it will fit together with the steps behind it and in front of it during this flattening.

- The *handrail:* In addition to rotating the main chain loops, the motor in an escalator also moves the handrails. A handrail is simply a rubber conveyor belt that is looped around a series of wheels. This belt is precisely configured so that it moves at exactly the same speed as the steps to give riders some stability.

- Power *disconnect:* Just as when operating around an elevator, we will shut down the mainline disconnect, which is located in the machine room or space. We will also perform lockout/tagout on the disconnect(s). The machine room or space for each escalator is located at the top of the individual unit. *Note:* Sometimes heavy-duty escalators have the driving machine located below the units, in a separate walk-in machine room.

In figure 27–9 an escalator truss under construction is shown. In figure 27–10, the same units are shown after the mall opened.

In figure 27–9, the escalator system services two upper levels. The machine room or space is located under the landing floor plates. The escalator shown would have a machine room or space for each escalator at each landing. In the photo shown in figure 27–10, there would be four machine rooms or spaces with disconnects in each for the individual unit.

Escalator installations also have safety circuit contacts built into the skirt panel at the landing in the event of a pile-up there. It is called the escalator skirt obstruction device. A pressure (amount depending on design) exerted against the skirt panel will cause the safety circuit to open, stopping the escalator. It must be reset by a key switch located at the bottom of the unit under the handrail on the newel skirt.

Fig. 27–9. Before

Fig. 27–11. Emergency stop switch

Fig. 27–10. After

Note the location of the emergency stop button (red) in figure 27–11, located in the upper right quadrant at *both* ends. Prior to 1983, it was located as part of the key switch located in the lower newel base. If an escalator accident is happening, any person should immediately reach for the emergency stop button and activate it. As you flip the protective plastic cover, a nuisance alarm will sound. This was developed to discourage vandals from activating the stop button.

Fig. 27–12. Mall escalator

Fig 27–13. Subway escalator accident

Fig 27–14. Closed machine space

In figure 27–12, showing an escalator located in a mall, the attention of most passengers is on everything and everyone, except the moving escalator. Most escalator accidents happen when the escalator is moving in the *down* direction. Add suitcases, a wider step, and more passengers, and theses are the makings of a rollback. A rollback can happen if an escalator is loaded with passengers beyond its rated capacity. The safety circuit is opened due to the overload, causing the escalator to stop, but the brake cannot hold it due to excessive overload, and it runs back down to the next landing, discarding passengers. What at first might appear to be a funny photo opportunity is a disaster, with passengers at the bottom of the pileup losing fingers, having the hair ripped off their scalp and suffering other severe traumatic injuries, including death.

The escalator in figure 27–13 was the scene of a fatality involving a young man who fell as the escalator was carrying him to train level in a transit station 140 feet below street level. Despite the actions of fellow passengers and responding firefighters, he could not be saved. His unfortunate death was the result of his necklace being caught in the step and the combplate.

Fig. 27–15. Open machine space

The machine room or space for an escalator is usually located as is shown in figures 27–14 and 27–15. It is usually located at the top of the unit, and it may be opened by either a tool specifically for that purpose or a large screwdriver. A sign can be seen on the wall inside the space, left by other mechanics, warning where the mainline disconnect switch is located. Many mechanics have been killed or injured while working on escalators, and most were caused by the failure to follow proper lockout/tagout procedures. At least one has been electrocuted while reaching to perform lockout/tagout and coming into contact with an energized piece of equipment.

of many accidents involving passengers young and old, where either clothing or fingers become caught in the machinery. A test to measure this gap has been developed called the Performance Index, or more commonly, the skirt index. As part of the ASME A17.1 Code, it is hoped that this will result in a decrease in the number of accidents. Another piece of equipment now being installed on escalators is the *skirt deflector,* which is a brushlike attachment applied to the skirt panel that will rub against a shoe or limb causing one to move what might be entrapped otherwise.

A new entry into the market is Otis Elevator's NextStep™ escalator (fig. 27–17). The escalator step actually has the skirt area as an integral part of each step. Ideally, it is thought that it will eliminate the side step entrapment, as there is no gap for anyone to get their foot or shoe or hand caught in.

Fig. 27–16.

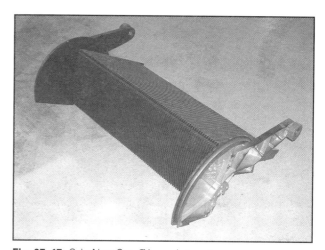

Fig. 27–17. Otis NextStep™ escalator step

Escalator Accidents

The child's foot in figure 27–18 is a simulation of how the space can be wide enough due to lack of maintenance or failure of a step adjustment to keep the space equal on both sides of the step. In figure 27–19, a child's finger is in danger of injury as she sits on the stair while riding with a parent or sibling on the escalator. Both of these scenarios would result in terrible injuries when the steps started down under, causing the crushing injury.

This poster is part of an industry-wide campaign to make riding escalators a safer experience (fig. 27–16). Part of the campaign is the development of features that will eliminate the gap that may exist between the side of the step and the skirt panel. This is referred to as a *side-of-step entrapment.* This space has become a source

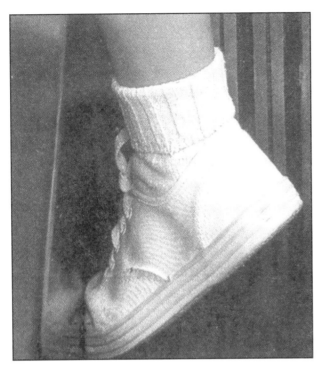

Fig. 27–18. Foot entrapment (courtesy of Carl White Associates)

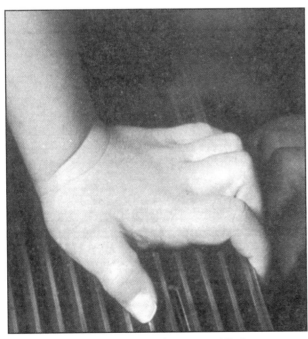

Fig. 27–19. Finger entrapment (courtesy of Carl White Associates)

The following summaries of government reports are reprinted with permission from the reports:

The Center to Protect Workers' Rights (www.cpwr.com)

Deaths and Injuries Involving Elevators or Escalators by Michael McCann, Revised March 2004[1]

- A 37-year-old man died from asphyxiation when his clothing became entrapped in the downward-moving steps and stationary bottom combplate of an escalator at a subway station. He was found, on his back, with the coat wrapped tightly around his chest, because part of the coat was dragged into the combplate. There were no witnesses as to how the coat became entangled (3/11/97, Washington, DC).

- A female, age 85, lost her balance and fell onto the escalator at a store. The cause of death was blunt impact to head, trunk, and extremities sustained in the fall (9/11/00, Richmond Heights, OH).

- A 12-year-old male was riding an escalator down (egress) from a baseball game when his right shoe got stuck between the stationary left side of the escalator. The victim sustained injury to his right big toe. The extent of the injury was not determined.(7/6/02, Anaheim, CA).

- A 5-year-old female was on the bottom step of a down escalator when her shoe got caught in the combplate. She reached down to get her shoe when her hand also got caught in the combplate. Her three middle fingers and part of her hand were amputated (2/19/03, St. Petersburg, FL).

- About 60 people were injured when the escalator they were riding down suddenly accelerated and they fell or were thrown at the bottom of the escalator (7/2/03, Denver, CO).

California Department of Health Services

FACE Report 01CA009

An Elevator Mechanic Helper Died When He Was Crushed in an Escalator While Performing Maintenance

Summary

A 37-year-old male elevator mechanic helper died when he was crushed in an escalator as he was performing maintenance. The victim had removed the escalator stairs and was standing inside the mechanism of the escalator when the power suddenly came on. The stairs began moving before the victim could get out and before the power could be turned off. There were no locks or tags on the controls that supply the electrical power to the escalator. *The disconnect switch at the circuit panel that fed power to the elevator had not been locked and tagged out* [italics added for emphasis]. The power came on when a coworker dropped the electrical circuit box, triggering a relay that started the escalator's movement. There was a mechanical blocking device on the escalator to stop movement during maintenance, but it was not used.

The CA/FACE investigator determined that, in order to prevent future occurrences, employers, as part of their Injury and Illness Prevention Program (IIPP) should:

- Ensure employees follow company policy and procedures on lockout/tagout.

- Ensure workers do not move electrical escalator equipment when all or part of someone is inside the escalator mechanism.

- Ensure employees block mechanisms from moving prior to performing repairs or maintenance.

The escalator system consists of the following major components:

- Landing

- Combplate

- Truss

- Tracks

- Reversal track

- Step

- Handrail

The main power disconnect switches are located in the machine space under the landing cover at the top of the unit, or in the walk-in machine room for a heavy unit. The landing covers can be opened either by a tool or by a large screwdriver. The walk-in machine room should be opened only by a service company door key.

Throughout the system, there are many automatic means for the system to shut itself down in the event that specific hazardous conditions exist. (See chapter 29, Escalator Safety Features.) Two of those features are the escalator skirt obstruction device, in which a design-set pressure, when exerted against the skirt panel, causes the safety circuit to be opened. The second comprises the emergency stop switches, which are located at the top and bottom of each escalator. In the past, they were installed low on the newel plate, at the bottom of the construction at floor level. The modern unit has them displayed high, in full view of the escalator and accessible to the public.

The entrapment history of escalators is not a pretty one, as seen in the photos with this chapter. Ideally, with the changes that have come about with the design improvements and a better educated riding public, some of these accidents can be prevented.

Summary

According to the latest available figures, there are 33,000 escalators operating in the United States and Canada, carrying approximately 245 million passengers per *day!* The ASME A17.1/CSAB44 Safety Code for Elevators and Escalators has set the standard speed for escalators at 100 ft/min, resulting in about 3,000 passengers per minute being transported.

Review Questions

1. How many escalators are there in the United States and Canada?

2. Name three of the six listed escalator parts.

3. Where will I find the mainline power disconnect switch?

4. What is the name of the new design in escalator steps?

Field Exercise

Arrange a visit to an escalator system in your inspection district. Determine the location of the machine space or room for that unit, and review the process for performing lockout/tagout with the elevator company representative on scene.

Endnotes

[1] McCann, M. *Deaths and Injuries Involving Elevators or Escalators.* Silver Spring, MD: The Center to Protect Workers' Rights, 2004. (See www.cpwr.com)

Chapter 28
Escalator Safety: Philosophy and Principles

The world of escalators has had its share of accidents and articles written about the deaths and injuries associated with them. In the Center to Protect Workers' Rights (CPWR) report of March, 2004, author Michael McCann, PhD, reported that in 1994 there were 7,300 injuries involving passengers on escalators.[1] In Singapore, Malaysia, the transit system (SBS) reported 72 accidents involving their escalators during a two-year period (Elenet-373 9/28/05). More recently, the Consumer Product Safety Commission estimated that there are 6,000 hospital emergency room-treated injuries associated with escalators each year (CPSC 2001). In this same report, approximately half of the sidewall-entrapment injuries involved children under age five years.

As with any incident, there are always a number of triggers that may have caused the incident. The first trigger may be the lack of proper maintenance on the unit, or a part breaking at a critical time. Another could be the way a person was riding on the escalator, paying attention to everyone and everything except where he was going and what he was doing. The under-five age group was particularly shown to suffer hand or foot injuries when becoming entangled with a combplate as a result of loose shoelaces or clothing getting caught. Then, there are all of the others reasons, including pile-ups, brake slippage, and the like.

Fig. 28–1. Transit system escalator down to train level (140 feet below street surface)

The Four Safety Principles

The members of the escalator industry, numerous individuals, and ASME A17.1 Safety Code for Elevators and Escalators have all worked together in making the escalator a safer ride. The skirt brush, skirt performance index measurement, radical new designs, and revitalized inspections are all aimed at preventing these serious and sometimes fatal accidents.

As with all incidents that we respond to, safety for all involved is our priority action. The following are the four safety principles:

1. *Prevent further movement of the escalator.* Primary action is power down, and lockout/tagout of the mainline disconnect. Driving wedges in between the steps and the side skirts will help further immobilize the unit.

2. *Guard the opening or scene of the entrapment.* Barricade the escalator at both ends to prevent people from trying to access the area. Shut down all access to the entire area, floor, or building, doing whatever is necessary to ensure safety and security of the work site.

3. *Protect the rescuers.* Provide proper protective clothing for the incident. If needed, have ground ladders thrown to enable access to all sides of the escalator.

4. *Protect the passengers.* Place a protective cover over entrapped victims to protect them from any shrapnel that may result from using tools to affect their release from the entrapment.

The Elevator Escalator Safety Foundation

We would like to make your fire service familiar with an organization that has created a number of programs aimed at preventing accidents by educating the riding public. Ideally, the information that follows will be of use to you in helping to protect the members of *your* public.

The Elevator Escalator Safety Foundation
362 Pinehill Drive, Mobile, AL 36606
info@eesf.org / 888-RIDE-SAFE

Mission: To educate the public on the safe and proper use of elevators, escalators and moving walks through informational programs.

United States Fact Sheet

Organization

- A recognized charitable organization in the United States and Canada.

- Chartered in the United States in October 1991 as a 501 (c)(3) charitable organization.

- Affiliated with Canadian charitable organization

- Affiliated with safety-minded organizations in the United Kingdom and Argentina

Statistical information (Source: *Elevator World,* June 1996)

In the United States,

- There are an estimated 600,000 elevators and 30,000 escalators.

- Over 90 billion riders travel on escalators each year.

- Over 120 billion riders travel on elevators each year.

- Over 575 million riders take an elevator or escalator in the United States daily.

Awards and Recognition

- Safe-T Rider© program was awarded the 1997 Youth Safety Award of Merit by the Youth Activities Division of the National Safety Council.

- First place Award of Excellence was given to Ripon Community Printers for "Be a Safe Rider" Coloring Book.

- Safe-T Rider program was listed on the Advance America Honor Roll for 1997 by the American Society of Association Executives Association for fifth straight year.

- Safe-T Rider program has been endorsed by the American Trauma Society.

- The EESF received 1996 Ad Wheel Award from the American Public Transit Association.

- U.S. Department of Transportation Secretary Federico Pena was named Safe-T Rider Safety Ambassador in 1995 and 1996.

- EESF is part of the Member Coalition for Injury Prevention and Control.

Foundation Programs

- Three active programs are available: (1) Safe-T Rider, (kits for schools and groups, www.safetrider.org); (2) National Elevator Escalator Awareness Week (second full week of November); and (3) A Safe Ride® (materials for adults, www.asaferide.org).

- Over 250 industry volunteers are involved.

Safe-T Rider Program

- Safe-T Rider is aimed at young children—one of the most at-risk groups.

- It was introduced in the United States in 1991 and in Canada in 1993.

- By June 2005, the program had reached nearly 4 million children parents and teachers since inception.

- It is currently active in 98 major U.S. and Canadian cities.

- Suggested implementation is in the fall to reach children prior to the holiday season.

- Message is reaching parents. The most frequent parent comment is, "I wasn't aware of that."

National Elevator Escalator Safety Awareness Week

- Celebration takes place second full week in November each year.

- Program introduced in 1994 to an audience of approximately 1.5 million in 10 U.S. cities.

- The next two celebrations expanded in virtually all locations to an audience of over 12.0 million riders.

- Detailed "How-To" Planning guide available free of charge for volunteers.

A Safe Ride

- A Safe Ride Program is meant for one of the most at-risk groups, older adults (seniors).

- The program was developed in cooperation with the renowned National Safety Council.

- The program consists of facilitator's guide, handbooks, and video. The handbook and video are available as stand-alone pieces.

- The program is available through the Foundation and free of charge to The National Council on Aging members and Senior Citizens Centers in limited quantities.

- There is a nominal charge for those not associated with The National Council on Aging or Senior Centers.

Escalator Safety Rules and Rationale

Unlike an elevator where the passenger-carrying compartment is fully enclosed, escalators and moving walks are only partially enclosed. The moving passenger is always adjacent to surfaces that are not moving, or possibly not moving. Precautions, similar to those exercised around any moving machinery, for example, subways or automobiles, should be observed. You are in control of your movement, and simple observations and advice can help you to travel as safely as possible.

Passengers only should be on escalators: No strollers, walkers, or carts of any type.

Use elevators instead of escalators when pushing a cart, stroller, or walker. Carrying these on an escalator will prevent you from holding the handrail and could cause you to lose your balance and fall. Additionally, they block your view when entering and exiting and could slow your exit, thereby delaying you and passengers riding behind from exiting safely and promptly.

Step on and off with care, and take extra care if you wear bifocals

Although escalators travel at a comfortable preset speed, boarding a moving step can be difficult to judge if you are in a hurry, carrying packages, or wearing bifocals. Care must be taken to promptly and firmly place your foot in the middle of a moving step while continuing to walk at a normal pace. Many escalators are equipped with a greenish light from within to help the rider determine the edges of the steps. Many escalator step edges are equipped with yellowish marks to help the riders likewise. If you miss the center of a stair—don't panic—simply adjust your feet.

Always stand facing forward and hold the handrail.

Standing facing the direction of travel allows passengers to maintain their balance and makes it easier to see the end of the ride and exit safely and promptly. Put your hand on the handrail as soon as possible when approaching an escalator or moving walk with the intent of riding it, and adjust your walking forward speed to that of the speed of the handrail. The reason that handrails extend beyond the moving steps or belt is so the riders may adjust their forward motion to that of the conveyance and not experience a sudden jolt when stepping on. It also places the rider's hand where it should be for the duration of the ride.

Pay close attention and attend small children; assist the elderly if they are hesitant.

Young children and the elderly are the ones that require extra care when riding escalators. Passengers need to help children on and off, making sure they stand up and keep away from the sides. Do not let them sit on the steps or play while riding. Have them hold the handrail or hold their hand if they cannot comfortably reach the handrail, while avoiding the side. Do not permit young children to ride on escalators or moving walks unattended.

Hold on to the handrail, but avoid the sides adjacent to the steps

Falls are the most common accident even on staircases. On escalators, the best way to prevent a fall is to firmly grasp the moving cushioned handrail. Both the steps and the handrail are moving at approximately the same speed. Passengers should avoid contact with the sides, because they are stationary.

Move quickly away from exit area.

The escalator and exiting passengers continue in motion, making it important to move quickly away from the exit area. Passengers should determine where they are going before they get to the exit area. Stopping to determine where to go in the exit area could make it difficult for passengers behind to exit and avoid those waiting.

On moving walks, stationary passengers should stay to the right and let walking passengers pass on the left.

Moving walks are designed to move passengers who are stationary as well as those who want to walk. Standing passengers should stay to the right and allow walkers to pass on the left.

More Escalator Rules and Advice

When entering an escalator:

- Watch the direction of the moving step.
- Step on and off with extra care. Take care if you are wearing bifocals.
- Hold children or small packages firmly in one hand.
- Grasp the handrail as you step promptly onto the moving step.
- Keep loose clothing clear of steps and sides.

When riding escalators:

- Keep to the right and face forward.
- Keep a firm grip on the handrail.
- Reposition your hand if the handrail moves ahead or behind the steps.
- Don't rest your handbag or parcels on the handrail.
- Don't window-shop while riding.
- Don't lean against the sides.

When exiting from escalators:

- Don't hesitate; step off promptly.
- Move clear of the escalator exit area; don't stop to talk or look around
- Be aware that other passengers may be behind you.

General tips:

- Keep your children's feet away from the sides of the escalator. Their shoes could be pulled in. Make sure your children's feet are in the center of the step.
- Never allow your children to sit on the steps. Should you witness another person allowing

this, please stop them. Never let your children play on the escalator.

- Make sure your children hold the handrail. As long as your children can grasp the handrail without their feet touching the sides of the escalator, it is safe. Holding the handrail will help ensure that they keep their balance and do not fall.

- If they are too small to hold the handrail without their feet being too close to the side of the escalator, they should only ride holding your hand.

- It is much safer for everyone if you take the elevator when you're using strollers, luggage, shopping carts, or you are overloaded with packages.

- Passengers must be careful of loose clothing, untied and long shoelaces, high heels, long hair, jewelry, and the like. These items can easily be caught in the combplate at the top and bottom. Only safe way to ride is standing up, facing forward, and holding the handrail with feet away from the sides.

- Don't rest your handbag, parcels, or body weight on the handrail. Putting too much weight on the rail may slow it down and pull you off balance. Hold on to the handrail gently until you are safely off the escalator. This is especially important for elderly people, whose sense of balance and depth perception is not strong.

Summary

The safety of the riding passenger is the most important aspect of all moving conveyances. Unfortunately, as we can clearly see from the Center to Protect Workers' Rights (CPWR) report quoted within this chapter, there are deaths and injuries are associated with both elevators and escalators each year. The report spells out in detail the causes of these deaths and injuries; however, many of them are preventable if workers follow their own industry requirements. Others are accidents that no one could foresee, with a combination of two or more remotely connected occurrences leading to the unfortunate end result. Some would call that Murphy's

Law, that whatever can go wrong will go wrong, and it will happen while you are working. Unfortunately, problems in design can be rectified only after the flaw has been the scene of an accident that has brought it to the attention of the authorities and designers.

The role of the Elevator and Escalator Safety Foundation (EESF) is an outstanding example of how educators, safety experts, industry representatives, and escalator designers have worked together to create an excellent resource for the asking. The various programs listed in this chapter are aimed at different age groups that have proven to be vulnerable to these hazards.

By following the safety principles here and in chapter 11, we can ensure that when operating at these incidents we will not become another number in the next CPWR report.

Review Questions

1. How many visits to the emergency room for escalator-related injuries are reported in this chapter?

2. What age group suffers half the side entrapment injuries involving children?

3. What are the four rules of safety that we follow from chapter 11?

4. What does EESF stand for?

5. List at least five of the safety rules and rationales.

Field Exercise

Visit the local elementary school(s) in your company area of responsibility and orient the administration to the EESF Safe-T Rider program.

Endnotes

[1] McCann, M. *Deaths and Injuries Involving Elevators or Escalators.* Silver Springs, MD: The Center to Protect Workers' Rights, 2004. (See www.cpwr.com)

Chapter 29
Escalator Safety Features

In the elevator/escalator world, there is no segment of the industry that receives as much respect and obedience to the code as that of the EPD, which stands for *electrical protective device*, the circuitry that electronically protects the system and its passengers from conditions that violate the ASME A17.1/CSA B44 Safety Code.

We have reviewed publications that list the EPDs that *automatically* protect the escalator and its passengers, and we list them in this section. There are other features listed in these publications, but this is the group that operates automatically to prevent injuries. Most if not all of them will go unseen and unrecognized by the firefighter, but will be performing their safety function during their operating life span. They have been added to the code one by one over the years, resulting from accidents or observations by concerned elevator and escalator people who wanted to prevent a recurrence of a dangerous condition. Many times these conditions resulted in injuries ranging from cuts and bruises to traumatic amputations and deaths. A full explanation of these definitions is in the ASME A17.1/CSA B44 Handbook-2004, section 6.1.6.3, Automatic Operated Safety Devices for Escalators and Moving Walks.

Electrical Protective Devices

Emergency stop button

The location gives the user of the emergency stop button a clear, unobstructed view of the escalator. The cover serves a twofold purpose, to prevent accidental contact and to discourage any unauthorized or unnecessary use. In the past, they were located under the handrail at the newel. This was very hard to see or reach. They are now located as shown in figure 29–1.

Fig. 29–1.
Emergency stop switch with vandal cover lifted

Speed governor

When an AC induction motor exceeds synchronous speed (an amount equal to its design slip), a braking action takes place electrically that prevents overspeeding.

Drive-chain device

The main drive shaft brake is created when the escalator machine is connected to the main drive shaft by a chain.

Stop switch in machinery spaces

The stop switch or mainline disconnect switch ensures that an escalator cannot be turned on by the start switch while a mechanic is working in the machinery space.

Escalator skirt obstruction device

In the event that a shoe, sneaker, finger, foot, and so forth, became entrapped between the skirt and the step, the skirt obstruction device would act to stop the escalator.

Escalator egress restriction device

An up-running escalator would bunch people at a closed rolling shutter or a closed door at the egress end, if the escalator were allowed to continue running. The egress restriction device prevents this.

Step upthrust device

The step upthrust device is required at the lower end, because an object trapped between the two steps would cause the lower step to resist transition to level, causing the device to operate and stop the escalator.

Disconnected motor safety device

This requirement is essential when a nonpositive drive connects the escalator motor to the machine.

Step level device

A downward displaced step entering the combplate, as a result of a missing wheel, its tire, or a failure of the end frame where the wheel fastens to the steps or pallet, can be hazardous.

Handrail entry device

Objects caught in the area where the handrail reenters the balustrade can produce bodily injury and/or damage to the escalator equipment.

Comb-step impact device

This requirement is intended to detect the first step to impact the comb and stop the escalator as a direct result of the impact.

Step lateral displacement device

This device is used in curved escalators only and will operate if a step is displaced horizontally.

Dynamic skirt panel obstruction device

This device minimizes the risk of entrapment between the dynamic skirt panel and the dynamic skirt panel cover.

Handrail speed monitoring device

There is a need to stop an escalator if the speed of the handrail deviates by 15% or more; this device monitors the speed.

Missing step and missing dynamic skirt devices

This requirement will prevent an escalator from continuing to run if a step is missing. It will prevent a void from a missing step to emerge from under the comb.

Tandem operation

If two or more units operate in tandem and the traffic flow will allow bunching if the unit carrying passengers away stops, they are electrically interlocked to stop the feeding unit if the unit carrying passengers away stops.

Step demarcation lights

Step demarcation lights warn the passengers that they are approaching the area where they must enter or exit the escalator.

Escalator smoke detectors

Smoke detectors are not required in wellways by the ASME A17.1/CSA B44 Safety Code for New Elevators and Escalators. If required by other codes or standards, such as NFPA 72, they must meet requirements to stop the escalator safely.

While the safety features that we have just listed are not always visible to the naked eye, other features are. The example shown in figures 29–2 and 29–3 is a skirt deflector installed on an existing escalator. This is a brush that is fastened to the skirt panel above the nosing. Through its presence and its positioning, it is hoped to prevent many entrapments by keeping the passenger's sneaker or sandal away from the edge of the step. This area was the scene of many side-of-step entrapments which resulted in serious injuries, primarily to children.

Fig. 29–2. Skirt panel deflector

Fig. 29–3. Escalator with skirt panel deflector

Summary

As we have seen throughout this book, the role of automatic protection plays an extremely important role in the world of elevators and escalators. The electrical protective device (EPD) provides automatic monitoring to more than 20 safety devices on the escalator and nearly as many to the moving walk. This system of protection has been developed through a number of means, including the following:

- Design changes as a consequence of problems that occurred
- Changes developed from accident histories
- Observations of change needs from within the industry

Two of the safety features that are related directly to the passenger in an entrapment are the emergency stop button and the escalator skirt obstruction device. The stop button can play a critical role in event of an accident in which the public can act to help by quickly pushing the button. The other feature, although automatic, would come into play when pressure was exerted against the skirt such as an entrapment would exert.

The skirt deflector brush and the skirt indexing are also part of the effort to make riding these units safer.

Review Questions

1. What is the EPD?
2. List the three ways change has come about in escalator safety
3. How many automatic EPDs are there in the escalator system?
4. Define a skirt deflector and how it is used.

Field Exercise

Locate and list three escalator systems in your inspection area. Contact the building owners and arrange a site visit with the responsible elevator company that is currently servicing the unit(s).

Chapter 30
Escalator Rescue Tools and Equipment

The tools and equipment necessary to effectively work on an entrapment involving an escalator are basically the same equipment that we would use in an elevator incident (fig. 30–1). The major difference is that with an escalator accident, we will not be hanging into an abyss of 1 to 30 or more stories. Most escalator accidents happen at the interface of the combplate and the landing. Nearly all of these happen when the escalator is in the down direction. This is not to say that we will not run into problems of height, as some of the designs for escalators have significant vertical distances to the ground. The example shown in figure 30–2 would result in one side of access being denied due to the construction.

Fig. 30–2. Courtesy of KONE Corporation

The lead complication in the incident will be the serious crush injury that may be entailed, and the crowded circumstances involving the patient and the rescuers. In earlier chapters, we have covered the causes of these accidents and the construction of the units. The first and most important action to perform is to power down, by removing power from the mainline disconnect. This must be followed by performing lockout/tagout for the unit(s) we will be working on, as well as the

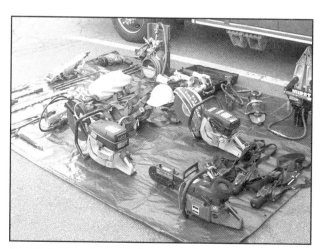

Fig. 30–1. Assorted tools for extrication

adjacent unit. If the unit is jammed to a stop from the accident, or someone has pushed the emergency stop button, we must make sure that the power is shut off to the escalators. This will be accomplished by opening the floor landing covers at the top landing for that unit, then finding the mainline power disconnect and turning it off. By shutting down both escalators, we will have additional work space for our activities, onlookers will be prevented from piling into our members and the mechanics whom we have called to the scene to assist us (fig. 30–3). The tools for this part of the operation may be as simple as a screwdriver or socket wrench set to open the fasteners for the cover (fig. 30–4).

The cover can then be removed for lockout/tagout, and then replaced to allow us more work space if it is a combplate entrapment. The combplate entrapment is usually relieved by removing the combplate screws, allowing the pressure to be taken off the foot or toe caught under them. Remove enough of the sections, and the problem is resolved. Sometimes the combplate entrapment can be very complex, due to the jamming of the person's limb down under the girder and under the step.

If the accident is a side-of-step entrapment, then the issue becomes more complicated. The irons (Halligan and flat head) will be needed to try to force the step away from the limb trapped (fig. 30–5). Screwdrivers or crowbars will be needed to remove sections of the skirt involved to relieve the pressure from behind the crush site. Long pry bars can be used to establish a bite or purchase, enabling wedges and other support materials, such as a small air bag, to be inserted to maintain the space and relieve the pressure (fig. 30–6).

Some of the other small tools that would be very effective in this limited space include the rabbit tool and the J bar that accompanies it, to gain a bite for the main lip of the tool (fig. 30–7). When using any of the pressure-inducing equipment, it must be kept in mind that the step itself may shatter, so patient care is even more pronounced in the use of these tools. Cover the patient and keep him or her covered to avoid shrapnel injuries.

Fig. 30–3. Open machine space

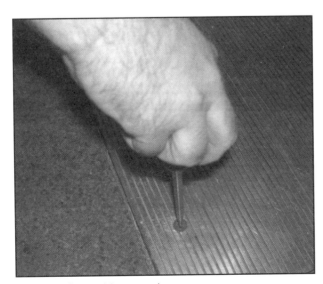

Fig. 30–4. Screwdriver opening cover

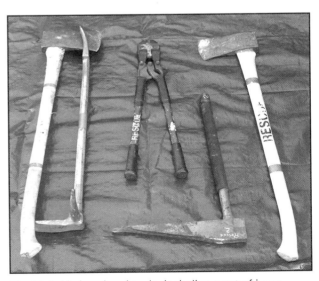

Fig. 30–5. Various hand tools, including a set of irons

Fig. 30–6. Small air bag unit

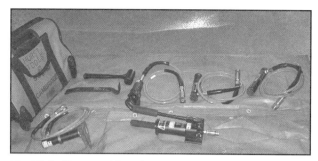

Fig. 30–8. Hurst tool-in-a-bag

At today's entrapment injuries, paramedics are essential (fig. 30–9). Both of us have seen the great transition from basic first aid to life-saving procedures that are being performed at incidents today. In the past, the patient was lucky if a responding fire department had bandages and a bag or valve mask for resuscitation. Today, there is a wonderful difference, because after we have saved a life in a fire, we can actually do something further in keeping that person alive, by administering drugs and IV fluids. This applies as well to the entrapment victim, as we can now simultaneously extricate and provide life support fluids and medicines.

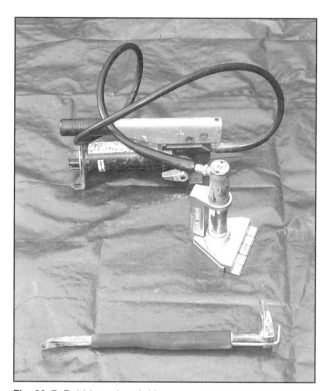

Fig. 30–7. Rabbit tool and J bar

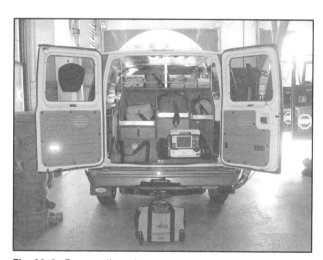

Fig. 30–9. Paramedic unit

Another tool that will work well in this close environment is the mini–hurst, or any other small edition of any of the various manufacturers of these tools (fig. 30–8).

In this chapter, we have established the types and uses of various hand and power tools that may be used in an escalator incident. Reviewing figure 30–1, it can be seen that there are a lot of choices to pick from. A tool that is golden today may be the worst choice tomorrow. Each incident will dictate what tool is best suited for that incident.

Summary

The tools and equipment that we will need and use on an escalator incident will basically be the same ones that we used during elevator incidents. The major points of entrapment are at the interface of the combplate and the landing, and nearly all in the down direction. The first and most important action is to power down the unit, followed by lockout/tagout procedures. Locating the cover to the machine space at the top of the escalator will enable the rescuers to open it via a screwdriver or small tool. The use of various small tools and power tools will provide the necessary power to force the entrapment away from the entrapped limb. The numbers and types of injuries can be disfiguring and range from cuts to amputation of fingers and other extremities. The major change in the response today is the ability to have paramedics on scene to provide the advanced care that will be needed once the patient is freed.

Review Questions:

1. What is the difference between elevator and escalator rescue tools?

2. How would you cope with a high deck escalator problem?

3. Describe the location of an escalator machine space?

4. What is the combplate interface?

5. Define the role paramedics play in these types of incidents.

Field Exercise

Organize a field exercise to a location such as a mall, transit center in your first due area, and pre-plan your response to the escalators at that location. Set up a field book entry so that other members of your fire company will benefit from this training.

Make Up!

To all firefighters, the term "make up," when heard on the fire ground, was a welcome sound. It meant pick up your hose, equipment, and the members of your company and return to quarters. The role of your fire company at this fire was done. On particularly brutal nights, in either subfreezing cold or wilting heat conditions, it was the company officer's job to get into the view of the chief running the fire. If he saw you, he couldn't forget you, so we all surmised. We would like to draw the same comparison to the work on this book, which has been assembled by both ourselves and our families. We have tried our best to give you the information that you will need as firefighters arriving on the scene of an elevator or escalator incident.

As with fighting any fire, managing elevator and escalator incidents requires an action plan, ongoing attention to safety, and teamwork. And let's not forget the resourcefulness of the elevator mechanic. The overarching reason for writing the book is to help firefighters become not only more knowledgeable about elevators and escalators but also more competent and confident to safely and successfully manage these types of incidents.

As we said in the introduction, we are not elevator (or escalator) people, but firefighters. There will likely be miscues and technical errors in this book, and for that we apologize. We have tried very diligently to keep those to a minimum, with a series of checks and counterchecks, to make sure we gave you the best effort possible.

This book could not have been done without the support of our families, friends, and the members of the Peer Review Group, who reviewed and commented on the chapters as we progressed to the end. We want to thank all who have contributed their time, effort, and support to the writing of this book. With all that said, now it is time to Make Up!

Returning to quarters,
John and Ted

Appendix A

Glossary of Elevator and Firefighting Terminology

This terminology originated in the Certified Elevator Technician (CET) Education Program, Course 1, *Introduction to Vertical Transportation*, Unit 1, Elevator History and Basic Safety and is used in part or whole with the permission of the copyright holder, the National Association of Elevator Contractors, 1298 Wellbrook Circle, NE, Suite A, Conyers, GA 30012.

Additional terms were added by the authors. These terms are noted with an asterisk (★). Some are from other sources and the authors composed others.

Terminology

AC (alternating current) A source of power for an elevator machine.

Acceleration A period during which the elevator moves at an ever-increasing rate of speed, usually referring to that period from stand still to full speed.

Alternated landing See Designated landing.

Annunciator An electrical device in the car that indicates visually to the attendant, by means of a target of light, the calls made by passengers waiting at the landings.

Arrival lights Arrival lights or hall signals are located on the outside of the elevator entranceway and illuminate to tell a passenger when the elevator has arrived.

Astragal The leading edge of a door panel made of a resilient material.

Attendant service operation An operation usually activated by an in-car key switch in an automatic elevator that allows the person in the car to control direction of travel, door closing, and car starting. The elevator will respond to both car and landing calls.

Balustrade The assembly of panels, newels, handrail, deck, and the like, which constitute the sides of an escalator or moving walk.

Barricade A guardrail system installed at the entrance of elevator hoistway or around escalator wellways at a construction site.

Barriers A temporary structure to restrict access to an escalator or elevator.

Brake An electromechanical device that prevents the elevator car from moving when the car is at rest, and no power is being applied to the hoist motor.

Buffer A spring or hydraulic device that is used to stop an elevator or counterweight from descending beyond its normal limit of travel by absorbing the kinetic energy. Spring buffers primarily store the energy while oil buffers dissipate most of the kinetic energy. Oil buffers are required for rated speeds over 200 ft/min.

Cable An electrical conductor containing several wires. An example is the traveling cable that provides electrical power and signal connections between the car and a fixed point in the hoistway.

Cam A shape structural, usually steel, that functions to operate switches as the car moves through the hoistway. It may be mounted on the hoistway or car. A retiring cam is usually located on the car and is moved by a motor to operate the switches. When the motor is deenergized, gravity causes it to drop and unlocks the hoistway door and operate door contacts.

Car, cab, and doors The enclosures that are built and mounted on a sling platform in which passengers and freight are carried. Doors are the movable portion of the car that protect the opening, providing access to the car.

Car, elevator The load-carrying unit including its platform, frame, enclosure, and door or gate. It is sometimes referred to as a cab.

Car enclosure The top and side wall of the car that is attached to the platform.

Car frame The supporting frame to which the car platform, upper and lower guides, and car safety and suspension ropes of the hydraulic plunger are attached.

Car operating panel (COP) The assembly of buttons and switches inside the elevator for operation and control.

Car platform The structure that forms the floor of the car and supports the load.

Certified elevator technician An individual who has met all of the application requirements of the NAEC CET™ program, has successfully completed all course requirements of the NAEC CET™ Level I and Level II, or has completed a documented program as approved by the NAEC Education Committee and passed the required qualifying examination.

Code The elevator code ASME A17.1 Safety Codes for Elevators and Escalators or ASME A17.3 Safety Code for Existing Elevators and Escalators. May also include other codes such as building codes and fire safety codes by reference.

Combplate The floor plate at the end of an escalator on which the comb teeth that mesh with the steps are mounted.

Compensating cables Steel cables that are attached to the bottom of the car and counterweight. They are used in tall buildings to help balance the car and the counterweight because of the extra weight caused by the hoist cables.

Competent person A person who is capable of identifying who is authorized to take prompt and corrective measures to eliminate existing and predictable hazards, on the job, in the surroundings or working conditions that are unsanitary, hazardous, or dangerous to employees.

Constant pressure door operation The need to hold a door open button until the door is fully opened. Release of the button before it is fully opened will reclose the door. Variations include constant pressure door close in which release of the button before it is fully closed will cause the door to reopen. The latter is a common operation with power-operated freight doors.

Contract speed The operating speed of an elevator, dumbwaiter, or escalator specified in the contract and/or specifications. This is the speed in feet per minute (ft/min) at which the unit should operate in the up direction with rated load (see Rated speed).

Control The system of governing the starting, stopping, direction of motion, acceleration, speed retardation, and leveling of the moving car for elevators and steps or pallets for escalators and moving walks.

Controller The electrical enclosure and components that perform the control functions.

Control solid state A control system in which control functions are performed by solid-state devices.

Counterweight 1. The structural steel frame containing weight connected by wire ropes to the elevator car to counterbalance the weight of the elevator. For a traction elevator, the counterweight is equal to the weight of the car plus a percentage of the rated load (often 40%). 2. Counterweights are also used on vertical car gates and other components.

Crosshead 1. The horizontal beam of a car frame on the top of an elevator car. 2. The part on a selector or floor controller member that travels in relation to movement of the elevator car.

Deflector sheave A grooved wheel that helps to direct the hoist ropes to the counterweight in the hoistway.

Designated landing The primary landing that the elevator will travel to when Phase I firefighters' service is initiated. If the sensor at the designated landing is activated, it will travel to the alternate landing.

Direct plunger driving machine A hydraulic elevator with the plunger or piston connected directly to the car frame or platform.

Do-not-start tag A company-approved tag used in conjunction with the company lockout/tagout program, advising the equipment has been deenergized for maintenance or repairs.

Door The structural unit of solid panels that open and close the hoistway and car entrances.

Door gibs★ Door guides, often referred to as gibs, normally have a plastic or metal member attached to the door panel that moves in the sill slot. During a fire, the plastic component of the gibs may melt out, causing a lack of stability at the bottom of the door panel. In the ASME A17.1 Safety Code, there is a requirement for a secondary safety retainer on the bottom of the door to ensure that the door cannot swing open from the bottom.

Door operator A motor-driven mechanical device that is mounted on the elevator car top, and opens and closes the elevator doors.

Door protection The means provided to protect entering and exiting passenger from moving doors.

Door-reopening device The device on an automatic door that senses an obstruction and changes the door motion by either stopping it, causing it to reverse or slowing it down. It may be a mechanical device or a noncontact device that uses photo cells or other means of detecting obstructions.

Door preopening The initiation of door opening through control circuits while the car is being stopped under normal operating conditions. The car is at rest or substantially level with the landing before the hoistway door is fully opened.

Door sill The threshold of the elevator cab door opening, which has grooves, that help to guide the bottom of the car door.

Door wedge tool★ A tool that is specifically designed to mechanically and temporarily lock and hold open a hoistway door.

Door track A steel guide on the elevator door opening on which the hanger roller or door guide runs.

Double-deck elevators Multicompartment elevators in a single shaft. Upper and lower decks are loaded simultaneously and serve odd- and even-numbered floors.

Driving machine The power unit that provides the moving force to move the elevator or other equipment. For elevators, it is the motor, brake, reduction gear, and drive sheave. For hydraulic elevators, it is the cylinder/plunger and connecting means.

Dumbwaiter Hoisting and lowering equipment including a car of limited size that moves in fixed guides and serves two or more landings. They do not carry passengers or operators.

Elevator personnel Persons who have been trained in the construction, maintenance, repair, inspection, or testing of equipment.

Elevator A hoisting and lowering mechanism, equipped with a car or platform, which moves in guide rails and serves two or more landings. The two primary classifications are freight and passenger. Freight elevators are used primarily for carrying freight and on which only the operator and persons necessary for loading and unloading are permitted to ride. Some other classifications of elevators are:

Traction elevator The car is counterweighted, and the friction between the traction sheave and rope lifts and lowers the elevator.

Electrohydraulic A direct plunger elevator whereby a liquid from a pump driven by an electric motor raises the elevator. Hydraulic elevators usually have a rated speed of less than 200 ft/min.

Roped hydraulic A hydraulic elevator that has its plunger or piston rod connected to the car by wire ropes, either directly or over sheaves.

Screw column The elevator car is supported, raised, and lowered by a lead screw and nut. Either the screw or nut may turn.

Private residence Is limited to a maximum rated load of 750 lb, a rated speed of 40 ft/min, and a rise of 50 ft and shall only be installed in a private residence or as access to a private residence.

Rooftop Opens onto an exterior roof and is limited to the roof and one landing below.

Winding drum Lifting is provided by winding a wire rope around a drum. Limited to use for freight elevators with speed of 50 ft/min and 40 ft of travel. (See the ASME A17 Code for additional classifications and limitations.)

Elevator building★ A building that contains one or more passenger or freight elevators.

Emergency door★ A door that is installed in the blind portion of a single blind hoistway at every third floor, but not more than 36 ft from sill to sill. The door provides access to elevator mechanics and firefighters during elevator emergencies.

Emergency light A special battery-operated lighting unit inside the elevator car that operates on loss of power to the elevator car. See the National Electric Code for definition of standby and emergency power.

Emergency exit An opening in the top and side of the elevator enclosure for use by maintenance or emergency personnel. The top panel may be opened only from outside the enclosure, and the side panel is hinged and requires a special tool or key to open from inside the car.

Emergency operation from a landing Various designations such as landing call emergency hospital service, medical service, EMT service, and so forth. Allows a designated elevator to be called to a landing for exclusive use by emergency personnel. Can also be designated "executive service."

Emergency stop switch A hand-operated switch in the car operating panel, which when thrown to the off position, stops the elevator and prohibits its running.

Escutcheon key★ See Hoistway door unlocking device key.

Escutcheon plate★ A metal protective plate that covers a hoistway door keyhole to prevent unauthorized use of devices other than those designed for that particular escutcheon.

Existing installation An installation that has been completed or is under construction prior to the effective date of the latest code.

Fall arrest system Personal fall arrest system means a system used to stop an employee in a fall from a working level. It consists of an anchorage, connectors, and a body harness and may include a lanyard, deceleration device, lifeline, or suitable combinations of these as defined by OSHA.

Fall hazard Exists when working more than 6 ft (1.8 m) above a lower level and an opening greater than 14 in. (254 mm).

Fall protection The use of guardrails, floor hole covers, or personal fall arrest systems when there is a potential fall over 6 ft (1.8 m).

Final limit The mechanically operated electric switch located in the hoistway set to operate and turn off power if the car travels beyond either terminal landing.

Firefighters' service Phase I—A special control that will take the elevators out of service and return them to a designated landing when a key switch at the designated landing is activated or a fire alarm initiating device in the lobby, machine room, or hoistway is activated. Phase II—Allows an elevator that has been placed on Phase I to be operated from inside of the car so that the person operating it has complete control. It is placed in Phase II by a key switch in the car.

Floor controller See Selector.

Foot protection Work shoes or boots that meet the company, industry, and/or ANSI Z41 standard.

Forcible exit★ The act of using a forcible entry tool or other forcing means to create or gain access to an exit or refuge area to escape a hostile environment. An example of forcible exit is the use of one or more forcible entry tools by firefighters trapped in an elevator car to breach a hoistway wall or top of car exit cover to escape a fire.

Full speed The contract speed at which the elevator should run.

Funicular Inclined elevator system using similar components to traction elevators.

Gate The car entrance on freight elevators that have open work. The size of the openings is restricted to reject a 2-in. ball. The door on the elevator car is often referred to as a gate even though it is a solid panel. In the early days, these were expanding collapsibles and therefore called gates. Solid units are normally called car doors but sometimes erroneously referred to as gates. Gates are no longer allowed on passenger elevators.

Geared driving machine A driving machine that uses a reduction gear between the motor and drive sheave or other means of transmitting power to the driven unit.

Gearless driving machine The motor, drive sheave, and brake are on a common shaft.

Governor A mechanical device connected to the elevator by a wire rope that monitors the speed of the car and counterweight and provides a signal to the control and activates the car or counterweight safety when an overspeed condition occurs. The governor is usually located in the machine room but may be in the hoistway.

Governor rope A wire rope that drives the governor and helps to slow the car down and stop it.

Governor tail sheave The governor tail sheave helps to guide the governor cable and keeps tension on it.

Ground fault circuit interrupter (GFCI) A device intended for the protection of personnel that functions to deenergize a circuit or portion thereof within an established period of time when a current to ground exceeds 4–6 mA.

Guide rail The structural member (usually T-shaped steel) fastened to the walls of a hoistway to guide the car and counterweight. The safety also clamps on the rail to stop the car or counterweight if an overspeed occurs. (*Note:* Some hydraulic elevators have round rails.)

Hoistway The opening through which the elevator travels. It is also referred to as the shaft.

Hoistway access switch A switch located at either or both the top and bottom landings to operate the car from the landing to gain access to the top of the car for inspection. When the switch is operated, the car will leave the landing in the proper direction with both the car and landing doors open. Some jurisdictions, known as "zoned access," require the distance the car will travel to be limited.

Hoistway doors The doors that, when open, allow access to the elevator hoistway.

Hoistway door unlocking device key★ A handheld device designed to manually unlock a hoistway door irrespective of the location of the elevator car.

Drop key★ A type of hoistway door unlocking device key having one or more hinged or knuckled joints.

Lunar key★ A type of hoistway door unlocking device key having an open cylindrical shaft with a lunar or half-moon shape at one end.

Tee key★ A type of hoistway door unlocking device key having a flat-based shaft with an inverted tee shape at one end.

Swing-door key★ A type of hoistway door unlocking device key having a solid cylindrical shape with an extended tip at one end.

Home landing A designated main terminal floor, usually the entrance floor to a building.

Hospital emergency service A special operation to provide complete control of an elevator to an operator for use during an emergency. There are several variations of this operation, and it is often considered a special variation of independent service.

Independent service Also called "hand operation," a special operation wherein a car is removed from automatic operation and no longer answers landing calls. It responds only to the calls registered on the car operating panel and is activated by a key switch in the car. Also used as a designation for "In Car Emergency Hospital" service.

Inspection Operation of a car from inspection switches located, under present A17 Code, on top of the car. Early versions used up and down buttons in the car to operate. Car is usually limited in speed to not more than 150 ft/min.

Interlock An electromechanical device that locks the hoistway door and completes an electric circuit to all operation when in the locked position. This device has two separate functions: 1. Prevent operation of the driving machine unless the hoistway door is locked. 2. Prevent the opening of the hoistway door from the landing side unless the car is in the landing zone and is either stopped or being stopped.

Jack The plunger and cylinder assembly on a hydraulic elevator. This is the hydraulic elevator driving machine.

Landing zone A zone extending from a point 18 in. below a landing to a point 18 in. above the landing.

Leveling The system or act of leveling a moving elevator platform level with the landing.

Leveling zone The limited distance above or below a landing within which the leveling device is permitted to cause movement of the car toward the landing.

Linear drive machine A linear electric motor mounted in the hoistway to move the elevator.

Lockout★ The placement of a lockout device on an energy-isolating device, in accordance with an established procedure, ensuring that the energy isolating device and the equipment being controlled cannot be operated until the lockout device is removed. Complies with OSHA. See Tagout.

Machine room The room that houses the power machinery for operation of the elevator.

Machine room-less (MRL) elevator An elevator in which all of the mechanical drive equipment is located within the hoistway, usually attached to the side wall, under the guide rail, or at the top of the hoistway, directly over the car, or in the pit, directly under the car

Mechanic, elevator See Technician, elevator

Motor generator set A unit consisting of an induction motor directly connected to a DC generator often on a common shaft. These units provide power for DC hoist motors.

Multicompartment elevator An elevator with two compartments, one over the other, that usually serves two floors at the same time.

Newel The post at the bottom and top of a flight of stairs that supports the handrails. In regard to escalators and moving walks, this would be the curved end around which the handrail turns.

Nonselective door operation Where both the front and rear entrance will open if a button for that landing is operated.

Nonstop switch A switch in the elevator car that when activated will prevent stopping for landing calls. These were used for authorized attendants and are seldom seen today.

Oil reservoir tank The oil reservoir tank is where the oil or hydraulic fluid used to move the elevators is stored.

Operation The method of actuating the controls divided into major categories summarized as follows:

Operation, automatic The starting of the elevator car in response to the momentary actuation of operating devices at the landing and/or of operating devices in the car or at a landing. Once the device is actuated, the car will stop automatically at the landing, and the doors will open. Another call will cause the doors to close, after a predetermined time, and the car will respond to another call.

Group automatic operation Automatic operation of two or more elevators that is coordinated by a supervisory control system including automatic dispatching that dispatches cars in a regular manner. It includes one button in each car for each landing served and up and down buttons at each landing (except terminal landings which have a single button). The stops set up by the car buttons are made automatically in succession as the car reaches the corresponding landing irrespective of the direction of travel or the sequence of actuation. The stops set up by the landing buttons are made by the first elevator in the group to reach the landing in the corresponding direction of travel.

Selective collective automatic operation One button in each car for each landing served and up and down buttons at each landing (terminal landings have only one button). Landing calls are answered as the car reaches the landing in the corresponding direction of travel; car calls are answered as the car reaches the landing irrespective of the sequence of actuation.

Nonselective collective automatic operation One button in the car for each landing served and one button at each landing. All stops registered by a car or landing button are made as the landings are reached irrespective of the number of buttons actuated, the sequence of actuation or the direction of travel.

Single automatic operation One button in the car for each landing served and one button at each landing. When a car or landing button has been actuated, the actuation of another car or landing button will have no effect until the response to the first button actuated is completed.

Car switch operation The movement and direction of travel of the car are directly and solely under the control of the attendant by means of a manually operated car switch or of a constant-pressure button.

Car switch automatic floor stop The stop is initiated by the attendant with a definite reference to the landing, after which the slowing down and stopping of the elevator is automatically affected.

Constant pressure operation The button or switch must be manually maintained in the actuated position to start and continue movement.

Operation inspection Operation that requires constant pressure on the button and is used for diagnostics, maintenance, repair, adjustment, inspection, and rescue.

Operating speed in the down direction The speed at which a hydraulic elevator operates while going down with rated load in the car.

Panel vision A small glass window in a car or hoistway door that permits persons at a landing to see when the car is at the floor or persons in the car see the floor at which the car is stopped. The maximum size and type of glazing are specified in the A17 Code.

Parking A feature incorporated into the signal system of an elevator or elevators by which an elevator receives a signal to always return to a preselected landing after all its car or landing signals have been answered and canceled. It is also called home landing.

Parking device An electrical or mechanical device that will permit the opening of the hoistway door from the landing side only when the car is within the landing zone of that landing. The device may also be used to close the door.

Penthouse The machine room above the hoistway on traction elevators.

Personal protective equipment (PPE) Protective equipment for eyes, face, head, and extremities. Including but not limited to protective clothing, respirator devices, and protective shields and barriers.

Photoelectric eye (P.E. Eye) A light beam or beams that span an elevator door opening and, when interrupted, cause the doors to reopen.

Pit The portion of the hoistway extending from the lowest landing sill to the bottom of the hoistway.

Platform The floor assembly on an elevator on which the passenger stands and/or the load is carried.

Platen The steel plate that connects the plunger to the car bolster (car frame or safety plank) on direct plunger hydraulic elevators and holeless hydraulics.

Plunger The round tube closed on at least one end that, extending into and from the hydraulic cylinder, moves the hydraulic elevator.

Plunger, gripper★ A safety device that is designed to stop a falling hydraulic elevator by automatically gripping the plunger.

Poling★ A procedure using a pole that is used by firefighters to reach and unlock a hoistway door from a location directly above, below, or next to the hoistway door.

Poling across★ A procedure using a pole that is used by firefighters to reach across and unlock a hoistway door from an elevator next to the hoistway door.

Poling down★ A procedure using a pole that is used by firefighters to reach and unlock a hoistway door from a landing directly above the hoistway door.

Poling up★ A procedure using a pole that is used by firefighters to reach and unlock a hoistway door from a landing directly below the hoistway door. This procedure is no longer recognized and recommended.

Poling tool★ A round or flat pole used to reach and unlock a hoistway door.

Position indicator A device (usually a lighted panel or number) that indicates the position of the elevator. A position indicator is inside the car and usually at the main lobby as well.

Power down★ The removal of mainline power to an elevator or an escalator. This will include the performance of lockout/tagout procedures after the main line disconnect has been shut off.

Pump motor The pump motor powers the pump that draws oil from the reservoir tank and pushes it through the oil line to fill the cylinder which moves the jack, causing the elevator cab to move upwards.

Rack and pinion driving machine The power unit on the car with a pinion gear that engages a rack attached to the hoistway or structure and moves the elevator.

Rated load The load that the elevator is designed to lift at rated speed.

Rated speed The speed that an elevator, dumbwaiter, material lift, escalator moving walk, or inclined lift is designed to operate up with rated load. This should be the same as the contract speed. (For hydraulic elevators, also see Operating speed in the down direction.)

Restrictor A collapsible, electromechanical, or mechanical device installed on an elevator to prevent opening of a car door when the elevator is outside the landing zone.

Riot control Somewhat similar to firefighters' operation, but used instead to keep elevators from a main landing in a civil disturbance.

Roller or shoe guides Wheels that are attached to the top and bottom of the elevator car stiles. Roller guides run inside of the guide rails and help to steady and guide the elevator car through the hoistway.

Rope The wire rope used to operate the governor while supporting the elevator and counterweight. Various sizes, types, and construction are used.

Rope gripper★ A safety device that is designed to stop an elevator that is overspeeding in the up direction by automatically gripping the hoist ropes.

Rope hydraulic driving machine The piston or plunger of a hydraulic unit that is connected to the car with wire ropes and sheaves. It includes the cylinder, piston or plunger, and sheaves and their guides.

Runby 1. **Bottom elevator** The distance between the car buffer and buffer strike plate when the car is level with the bottom landing. 2. **Bottom counterweight** The distance between the counterweight buffer and buffer strike plate when the car is level with the top landing. 3. **Top runby hydraulic elevator** The distance that the elevator can move above the top landing before the plunger stop ring engages the top of the cylinder.

Safety edge See Safety shoe.

Safety factor (or factor of safety) The ultimate strength divided by the maximum load or stress.

Safety (car or counterweight) The mechanical device usually located at the bottom of the car and counterweight that is activated by the governor to stop the motion of the car and counterweight in the event of an overspeed condition in the down direction.

Safety rope (tail rope) A short rope that connects the safety to the governor rope.

Safety circuit The circuit that, when opened by any safety device (such as a governor overspeed switch, in-car stop switch, etc.), will cause power to be removed from the driving machine motor and brake.

Safety shoe The leading edge of a power-operated horizontal sliding door that is mounted to the door panel with hinged linkage allowing it to move if obstructed. When obstruction occurs, a switch is operated that causes the door to stop and reopen. Sometimes called a safety edge.

Selective door operation Where front and rear entrances (or side entrance) are on an elevator, each will have a car button for the designated landings. Only the door for which the button is operated will open.

Selector (or floor controller) The electromechanical device that can be electrically or mechanically connected to the elevator car to accurately mimic its movement. It interfaces with the control system to provide information on and control the car position and provide information on the car's position to the controller.

Selector tape A small steel-grooved cable that is attached to the selector and the elevator car. As the car moves up and down the hoistway, this tape relays the car's movement to the selector, which in turn, informs the controller of the car's position.

Selector tape sheave A grooved wheel over which the selector tape travels.

Shunt trip★ An electric device that is designed to remove the mainline power from all elevators in a hoistway, on actuation of a heat detector located in the machine room protected by an automatic fire sprinkler system.

Site guard A vertical member mounted on the leading edge of the hoistway side of the hoistway door to reduce the distance to the car door and restrict access to the door clutch and the like.

Slack rope switch A switch that is mechanically arranged to detect a slack rope condition and cause the power to be removed from the driving machine motor and brake when slack rope occurs.

Solid state An electrical device or circuit that has no moving parts. It operates and controls current and/or voltage without mechanical contacts or movement.

Speed governor See Governor.

Spirator (door closer) A spring-operated door closure that winds the cable up on a reel operated by a constant pressure spring.

Stack The stack or guide rails are an accurate, vertical line of rails located in the hoistway. The term *stack* may also be used when referring to conduit and electrical wiring.

Stop switch A manually operated device that removes power from an elevator or escalator driving machine motor and brake.

Stopping The action of final motion of the elevator from leveling speed to an accurate floor level.

Tagout★ The placement of a tagout device on an energy-isolating device, in accordance with an established procedure, to indicate that the energy-isolating device and the equipment being controlled may not be operated until the tagout device is removed. Complies with OSHA. See Lockout.

Technician, elevator An elevator technician is a person that has been trained in and is skilled in the construction, modernization, repair, and/or maintenance of elevator equipment. They are sometimes referred to as elevator constructors or mechanics.

Terminal landing The top and bottom landings.

Terminal stopping device A device, independent of the operating device, located at or near a terminal landing, whose purpose is to slow down and stop the elevator or dumbwaiter.

Toe guard (car apron)★ A physical barrier that hangs down a maximum of 48 in. from the front of the bottom of the car assembly. If the car were up in the opening, it would create a barrier to the open hoistway from the public.

Top of car inspection (TOCI) A device on top of the car required by ASME A17.1 Code that permits elevator personnel to operate the elevator from the car top.

Traction machine An electric machine in which the friction between the ropes and drive sheave is used to move the elevator car.

Traction sheave A grooved wheel over which the elevator hoist cables travel. It is the friction between the hoist cable and the traction sheave that helps to move the elevator cab up and down the hoistway.

Travel (rise) The vertical distance between the bottom and top terminal landings of an elevator, dumbwaiter, or escalator.

Traveling cable The electrical wiring that carries electrical power, signals, and communication between the elevator car and the hoistway. It consists of flexible stranded and insulated wires bundled together and enclosed in a sheath.

Unintended motion Any movement of the elevator that is not caused by the control system.

Valves Valves are what open and close to allow the flow of oil to and from the hydraulic cylinder to move the elevator cab up and down.

Winding drum driving machine A geared drive machine that winds the rope on a drum to lift and lower the elevator car.

Zone, landing★ A zone extending from a point 18 in. below a landing to a point 18 in. above the landing.

Zone, unlocking★ A zone extending from the landing floor level to a point not less than 3 in. nor more than 18 in. above and below the landing.

Bibliography

A17.4 Guide for Emergency Personnel-1999. New York: American Society of Mechanical Engineers.

Bunker, M. "Interfacing Elevators and Fire Alarm Systems." Rolf Jensen Associates. necdigest, Fall 2003.

Coan, S. "Sprinklers in Elevator Machine Rooms, Elevator Hoistways and Elevator Pits." State Fire Marshal, Commonwealth of Massachusetts, March 30, 2005.

Donoghue, E.A. *ASME A17.1/CSA B44 Handbook–2004 Edition.* New York: CPCA American Society of Mechanical Engineers, 2004.

"Door Restrictors." Otis Elevator Company, 1998, MOD-9506(0998).

"Elevators and Fire." The Symposium on Elevators and Fire, February 19–20, 1991. Baltimore, MD. New York: ASME.

"Elevators, Fire and Accessibility." The 2nd Symposium on Elevators, Fire and Accessibility April 19–21, 1995. Baltimore, MD. New York: ASME.

Emergency Evacuation Elevator Systems Guidelines. Chicago: Council on Tall Buildings and Urban Habitat, 2004.

Emergencykey.com PO Box 11024 Lexington, Kentucky 40512–1024.

"FDNY Training Bulletin–Emergencies 1–March 15, 1997, DCN: 3.02.17–Elevator Operations." Fire Department of the City of New York.

The Firefighter's Handbook: Essentials of Firefighting and Emergency Response. Albany, N.Y.: Delmar, 1999.

Goodwin, J. *Otis: Giving Rise to the Modern City.* Chicago: Ivan R. Dee, 2001.

Gray, L. "A Brief History of Residential Elevators: Part 1." *Elevator World* (January 2005), p. 110.

Gray, L. "A Brief History of Residential Elevators: Part 3." *Elevator World* (March 2005), p. 142.

Gustin, B. "Mechanical Elevator Door Restrictors: What firefighters need to know." *Fire Engineering* (11/7/2003 download).

Halligan, H.A. "The Halligan Tool." New York: WNYF 1st/97.

"High Rise Building Evacuation Plan." *DRAFT* FDNY Training Division Fire Department of the City of New York.

Jarboe, Ted., Elevator Rescue, compiled training notes (notebook), 1980.

Kidd, J.S., and Czajkowski, J.D. *Elevator Rescue Manual.* First Due-Rescue Company series. St. Lous: Mosbey-Year.

Koshak, J. "Falling Hydraulic Elevators in Need of a Safety." *Elevator World* (December 2001), p. 101.

McCain, Z. "Elevators 101." *Elevator World* (2004).

McCann, M. *Deaths and Injuries Involving Elevators or Escalators.* Silver Springs, MD: The Center to Protect Workers' Rights, 2004. (See www.cpwr.com).

"Multiple Fatality High-Rise condominium Fire." Clearwater, Florida. USFA-TR-148/June 2002.

"Sprinkler Systems and Fire alarms for Elevator Machinery Rooms, Hoistways, and Pits." Seattle Fire Department Administrative Rule 9.08.05–4/21/05.

Stenqvist, J. "Sprinklers and Elevators: Can They Coexist?" *CIPE PM Engineering.*

Swerrie, D.A. "Another Viewpoint Column." *Elevator World* (September 1999), p. 80.

Witham, D. "Passenger Elevator Door Operators 101." *Elevator World* (April 2002), p. 106.

Index

C

Index

I

Illinois Fireman's Rule, 182–183
improvised hoistway door release tools, 123
In the Mouth of the Dragon (Wallace), 79
incident commander (IC), 4, 36, 95, 98, 138, 196, 199
independent service, 44, 249
inspection, 43, 68, 123
    box, 202–203
    operation, 68, 249, 251
    station, 7, 10, 25, 138, 148
    switch, 94, 156, 159, 249
    top of car, 7, 25, 253
interlock, 26, 28–29, 106–107, 249
    location of, 28, 33
International Code Council, xv
International Fire Service Training Association
    (IFSTA), xv, 191
irons, 144, 150, 242

J

J bar, 148, 242
jack, 249, 251
    stand, 51
    telescopic, 54–55
    vertical, 53
JarClose elevator tool (pole), 128
jump bag, 103, 160
jump pit, 9

K

keeper, 28–29
Kevlar, 68
key
    box, 109
    damage, 124
    drop, 102, 109–113, 124–125
    escutcheon, 105
    FEO-K1, 191–193
    Firefighters' Emergency Operation, FEO-K1,
        191–193
    firefighters' service, 102, 198

    hoistway door, 12, 26, 101–102, 105, 108–109, 111,
        123–125, 249
    hook (extrusion), 119–120
    lunar (half-moon), 102, 108–109, 114–115
    Phase I, 40, 42, 197
    swing door, 118, 124
    tee, 102, 109, 115–117, 124–125
keypad, 38
kinetic energy (KE), 23
Knox-box, 12, 102
KONE Corporation, xvi
KONE Mono-Space machine room-less (MRL)
    elevator, 71–72, 79
KONE PM traction machine, 15

L

ladders, 103
landing
    alternate, 67, 198, 247
    designated, 39–41, 65, 177, 178, 198, 247–249, 252
    emergency operation from, 248
    escalator, 222
    home, 249, 251
    terminal, 248, 250, 253
    zone, 59, 249, 251, 253
Lapierre, Ray, 221
level
    alternate, 67, 198, 247
    designated, 39–41, 65, 177, 178, 198, 247–249, 252
    of training, 208
leveling, 107, 246, 249, 253
leveling zone, 249
lifeJacket, 76–77, 79
light-emitting diode (LED), 3, 23–24
limited use/limited access (LULA) elevator, 81, 85–86
linear drive machine, 250
lockout, 97, 250
lockout device, 97
lockout/tagout, 96–97, 99
lockout/tagout set, 12
lunar (half-moon) key, 102, 108–109, 114–115

M

N

O

Index

P

panel vision, 251
panels
  fast, 21
  front, 21
  insert, 21
  slow, 21
paramedic unit, 243
parking, 251
parking device, 251
passenger entrapment, 153–155, 161–162
  documentation of, 161
  emergency guidelines, 159–162
  nonemergency guidelines, 155–159
penthouse, 12, 49, 68, 251
permanent magnet (PM) motor, 68, 73–74
permanent magnet synchronous motor (PMSM), 72
personal protective equipment (PPE)
  definition of, 196, 251
  use of, 92, 99, 133, 169, 196, 211
Phase I key, 40, 42, 197
Phase II panel, 45, 47
photoelectric eye (P.E. Eye), 23–24, 251
pickup roller assembly, 26, 28
pike pole, 128
pinned-victim incidents, 165–166
  emergency care in, 166
  procedures for, 167–168
  technical rescue team and, 166–167
pit, 251
  door, 9
  jump, 9
platen, 251
platform
  area of, 85
  definition of, 246, 251
  use of, 83, 210, 246–247, 249
plunger, 51
  cylinder system, 167
  definition of, 251
  direct-, 52, 68, 247, 251
  follower guide, 55
  gripper, 75–76, 251
  hydraulic system, 53, 246–247, 249, 251–252
  stop ring, 252

poling, 132, 251
  across, 129–131, 135
  down, 131, 133, 135–136
  open, 135
  passing, 133
  safety, 127, 133–134
  sequence of, 133
  tools, 12, 102, 128, 130–131, 251
  up, 129
polyurethane (PU) covering, 68–69
polyvinyl chloride (PVC), 76
portable radio, 12, 103
position indicator, 39, 154, 251
power
  control, 229
  disconnects, 14, 18
  down, 125, 251
  for driving machine, 44, 107, 252
  failure, 155–156
  oil, 45
  -operated doors, 60
  -operated gate, 60
  selector panel, emergency, 35
  water, 49
power take-off (PTO), 14
predetermined marooning, 188–189
pump, 14
  hydraulic, 52, 107, 155, 214, 247
  motor, 251
  room, 213–214
  submersible, 50
PVC (polyvinyl chloride) liner, 52

R

rabbit tool, 12, 147–148, 242–243
rack and pinion driving machine, 251
rated
  load, 85, 246, 248, 251
  speed, 82, 246–248, 251
reel closer, 26
release roller arm, 117, 122, 135
remote firefighter service, 36
remote firefighters' emergency operation, 35
Reno, Jesse, 217, 219
rescue rope and harness, 102–103, 148
rescuvator, GAL, 76, 79

S

Index